THE ARK'S ANNIVERSARY

The Ark's Anniversary

(to page 5)

GERALD DURRELL

Arcade Publishing • New York

LITTLE, BROWN AND COMPANY

3

First U. S. Edition 1991

Library of Congress Cataloging-in-Publication Data

Durrell, Gerald Malcolm, 1925–
   The ark's anniversary / Gerald Durrell.
      p.   cm.
   Includes index.
   ISBN 1-55970-140-4
   1. Jersey Zoological Park — History. 2. Durrell, Gerald Malcolm, 1925–
3. Zoologists — Great Britain — Biography. I. Title.
QL76.5G72T742   1991
591'.74'442341 — dc20   91-11698

Published in the United States by Arcade Publishing, Inc., New York,
a Little, Brown company

10 9 8 7 6 5 4 3 2 1

RRD-VA

Printed in the United States of America

THIS BOOK IS FOR

THOMAS LOVEJOY

WITH WHOSE HELP, HUMOUR AND HARD WORK
WE HAVE ACHIEVED MUCH

Contents

<pause>

Narrator's Note: Captions to photographs found in The print edition are located at The end of this recording: side ⑤.
     End of Narrator's Note.

<pause, Then to page 11 >

7

List of Illustrations

Foreword by H.R.H. The Princess Royal

My association with the Jersey Wildlife Preservation Trust began, as it has done for thousands of others, on a train with a book written by its founder, Gerald Durrell. He has the ability, shared with few other authors, to extravasate a spontaneous explosion of mirth that surprises the reader as much as it does the unsuspecting travelling companion. To eat one's sandwich while reading Mr Durrell is to court still greater embarrassment.

I realize now that he uses his own exclusive brand of anthropomorphism to build relationships – and therefore, inevitably, bonds – between his reader and other animal species. I also recognize that he employs this singular gift to reach a multiple audience, a strategy he continues to harness to great effect in other ways.

I have now visited the J.W.P.T. several times and while each is memorable in its own way, none stands out as being more significant than that of the Trust's 21st and the Zoo's 25th joint anniversary in 1984, when I was asked to open the International Training Centre for the Conservation and Captive Breeding of Endangered Species.

You will read that, via graduates of this Training centre, the Trust's influence and activity is spreading across the globe on a scale wholly disproportionate to its modest headquarters in the Channel Islands.

I believe that it behoves all of us, as guardians of the living world we have inherited, to see that we pass on this priceless inheritance to the next generation. In order to do that, however, we must understand why it is necessary and how to do it. One of the most exciting developments, growing out of a realization that captive breeding can, of itself, create an opportunity for learning, is the adaptation of the breeding centre to provide a focus for public education.

I am delighted that The Princess Royal Pavilion in Jersey will provide one such opportunity, so that over 350,000 visitors to our headquarters each year can be taught the philosophy behind the Trust's objectives. Of at least equal significance is the first Zoo Educator's Course which will teach zoo staff from developing countries how well maintained animal collections can be utilized to impart the principles of wildlife conservation.

Once again Gerald Durrell and his small team seem to have found a way to reach an international audience of barely estimable numbers.

To say that one person, or indeed one organization, cannot do it all, is a truism. But I feel bound to suggest that if everyone and every biologically based institution did as much as Mr Durrell and his Trust to help stitch and darn our planet's threadbare ecology, there might be fewer holes in our natural defences than there are today.

All Gerald Durrell's books are worth waiting for. This one is no exception and will, I hope, serve to convince many more people that where there is a will and a well thought out way, the impossible becomes commonplace and even miracles don't take quite so long.

Anne

Author's Note

Most authors complain of too little material. In this book I complain of too much, for I have been forced, by the exigencies of space, to leave out much that I would have loved to include. However, it has taught me the truth of the old adage that you cannot get a quart into a pint pot.

As the Trust has grown and prospered and now has supportive sister organizations both in the United States and in Canada, we have come to use the word 'Trust' as an all-encompassing description since, though separated by oceans and vast distances, our work, our objectives and our aspirations are the same. Therefore when I use the word 'Trust' in this book it covers not only the work of Jersey but of the USA and Canada as well.

A Word in Advance

I do not think it possible for many people at the age of six to be able to predict their future with any accuracy. However, at that age I felt confident enough to inform my mother that I intended to have my own zoo and moreover, I added magnanimously, I would give her a cottage in the grounds to live in. If my mother had been an American parent, she would probably have rushed me to the nearest psychiatrist; however, being fairly phlegmatic, she merely said she thought that it would be lovely and promptly forgot all about it. She should have been warned since, from the age of two, I had been filling matchboxes and my pockets with a wide variety of the smaller fauna that came my way, so the progress from a matchbox to a zoo could have been predicted. It is nice to record though that, before she died, I had fulfilled my promise and taken her to live in my zoo, not in a cottage but in a manor house.

Looking out of the windows of my first-floor flat in Les Augres Manor is an unpredictable operation and would give any psychiatrist pause for thought. From the living-room windows, for example, you are suddenly transfixed – while in the middle of pouring out a refill of pink gin for your guest – by the sight of the Przewalski horses running their variation of the Derby around their paddock, and you watch them breathless, wondering which muscular, pinky-brown animal is going to win. Meanwhile, with no explanation for your sudden inhospitable immobility, your guest is left, as it were, unquenched.

In the dining room worse befalls. Your carving is brought to a halt in mid-joint because you have let your gaze stray out of the window and have caught sight of the Crowned cranes doing their courtship dance. Their lanky legs are thrown out at the most unanatomical angles as they pirouette around like bedraggled, failed ballet dancers, leaping high in the air, dextrously juggling twigs as love tokens and uttering loud, rattling, bugle-like cries.

Modesty prevents me from relating what can be seen from the bathroom window when the Serval cats are loudly and apparently agonizingly in season, moaning their hearts out, screaming with love and lust. However, worse – far worse – befalls you in the kitchen should you lift your eyes from the stove and let them wander. You are confronted by a large cage full of Celebes apes, black and shiny as jet, with rubicund pink behinds shaped exactly like the hearts on valentine cards, and all of them indulging in an orgy which even the most avant-garde Roman would have considered both flamboyant and too near the knuckle. Close contemplation of such a spectacle can lead to disaster, such as irretrievably burning lunch for eight people as they arrive to partake of it. This happened to me on one occasion and I discovered that even friends of long standing do not take kindly to boiled eggs when they have been churning up their gastric juices in expectation of a five-course gourmet meal.

There are even worse things. One morning I was entertaining a coterie of extremely ancient and wobbly octogenarian conservationists, who were making steady inroads into my sweet sherry. I was just about to suggest (while they had what wits they possessed still about them) that we sallied forth to look at the animals, when I glanced out of the window and saw, to my horror, slouching through the forecourt among the spring flowers, Giles, our biggest, most hirsute and potentially lethal Orang utan. He looked like a gigantic walking hearthrug of orange and blond hair and he had that lurching side-to-side movement thought to be the prerogative of sailors who have spent many years on the bosom of the ocean and an equal number of years on the rum. I was trapped and for the next hour I had to ply my geriatric acquaintances with more and more sweet sherry and they got progressively more and more inebriated. At last the happy news came that Giles had been darted and returned to his rightful quarters, and I could get rid of my by now extremely convivial conservationists. But my blood ran cold at the thought of what would have happened if I had ushered them (all the worse for the application of the Demon Drink) out of the front door at the precise moment when Giles came shambling into the forecourt.

But why take on a zoo? – relatives and friends have asked me plaintively, and not a biscuit factory, or a market garden, or a farm, or something safe and respectable?

The first answer was that I never wanted to be safe and respectable; I could not imagine anything duller. Second, I did not think that the ambition to own one's own zoo was so outrageously eccentric that it warranted all your nearest and dearest looking at you as if you were ripe and ready to be fitted for your first – and probably last – straitjacket. To me the thing was perfectly straight-forward. I was deeply interested in all creatures that lived with me on the planet and wanted them at close quarters so that I could watch them and learn about and from them. What simpler way of accomplishing this than to found my own zoo?

In those palmy days, of course, I had no conception of the amount of money and hard work that would have to be put into such a project before the dream could become a reality, nor did I have any idea of the importance of zoos and what, ideally, they should be. Selfishly, I regarded a zoo as being a large collection of exotic animals gathered in one spot for my personal edification, nothing more. But as I grew older and nearer to my ambition I worked in zoos and collected animals around the world for them and I began to view them in a somewhat different light from the uncritical one I had until then adopted.

What I had in mind was an almost completely new concept of the motivation of a zoological garden. Its first major objective should be to act as an adjunct to the whole conservation movement by the setting up of viable breeding colonies of those endangered species whose numbers had dropped so drastically that they could no longer cope with the hazards of life in the wild. This should in no way be misconstrued (as it has been by some conservationists) as an action which merely confined these animals to captivity. The idea was that the captive colonies should be set up merely as a safe-guard against extinction, while at the same time the most stringent efforts should be made to preserve the wild habitat and wild popu-lations of the species concerned and to release back to the wild captive-bred animals when their habitat had been made safe. This, it seemed to me, was a zoo's major *raison d'être*.

Second, a zoo should help establish breeding colonies of the species in their countries of origin and train people from those countries in captive breeding and reintroduction techniques.

Third, a zoo should promote studies to learn more about the animals themselves, both in the wild and in captivity, and through this knowledge find better and more rapid ways of helping them to avoid vanishing from our world.

Last but definitely not least, a zoo should promote conserva-tion education both in the country where it is found and in the country from which its endangered species emanate, and generally where such education is most urgently needed.

I discovered to my dismay that a very high percentage of zoos were bad. They were bad because they had no real motivation and were run simply as showplaces. Any motivation there was con-sisted of trying to obtain 'box-office' animals to enhance the gate receipts. The animals were badly fed and badly caged for the most part, and the breeding results (if any) were poor and happened more by accident than by design. Very little scientific study was done on this vast array of species about which practically nothing was known and the attempts to educate the zoo-going public were pathetic at best.

I have written elsewhere (*The Stationary Ark*) that when Flor-ence Nightingale was faced with the appalling hospitals of her day she did not suggest closing them all down. Knowing that they had an important part to play, she suggested that they were made *better*.

I am in no way comparing myself with that redoubtable woman, but the same sort of situation existed (and exists still) with zoos. I felt the fact that – as institutions – they had fallen into disrepute was entirely their own fault. I believed that zoos could be important if they were run properly, excellent institutions for scientific research and education, but – above all in this day and age – centres of captive breeding to assist in the saving of endangered species.

So, quite simply, I wanted a zoo run on these lines, lines I thought every zoo should follow. I was not at all sure if it would work, but then the Wright brothers did not know if they could fly until they took to the air. So we tried, and now – years later, after a lot of hard work and many mistakes – we have proved that it can work. That is why this book is called *The Ark's Anniversary*, for we recently celebrated our twenty-fifth birthday. And this is the story of some of the many things which happened to us during our years of growing up.

1

The Emergence of the Manor

A t the age of twenty-one I inherited £3000, a princely sum but not sufficient to start a zoo. So I decided to become an animal collector *for* zoos. It was a short-lived career, because I discovered that what most dealers did was to cram twenty creatures into a cage designed for one and increase the price on the survivors. If they all survived, well and good. I could not indulge in this sort of slave traffic, so my cages were spacious and my animals well cared for: so I lost all my money. However, the experience proved invaluable. It gave me a wide schooling in the keeping of animals in the tropics, their illnesses and their quirks. It taught me that zoos were not all that I had thought them to be.

Then, penniless, at my elder brother's insistence I started to write. I was lucky. My first book was what they now call a smash hit and I have been lucky that all my subsequent books have been equally popular. With this change in my fortunes, my thoughts turned zoowards again. Borrowing £25,000 (against as yet unwritten masterpieces) from my kindly and long-suffering publisher, I decided to try to set up on the south coast of England, only to find that a succession of Labour governments had enmeshed the country in such a Kafka-like miasma of bureaucracy that the average citizen was bound immobile by red tape and it was impossible to get quite simple things agreed to by local government, let alone something as bizarre as a zoo. So, with introductions from my publisher, I went to Jersey – small, beautiful and self-governing – and within a few hours of landing I had found Les Augres Manor, and within forty-eight hours I had received the go-ahead. However, I had not rushed into the whole thing with ill-considered enthusiasm and without taking advice. I approached everyone I knew in what might loosely be called the biological zoo world whom I knew approved of the idea of captive breeding. The first was James Fisher, great ornithologist and ardent zoo man. He helped by telling me I was mad.

'You're mad, dear boy,' he said, staring at me from under his mop of iron-grey hair, looking like an extremely worried Old English sheepdog, 'quite mad. I must really advise you against the Channel Islands.'

He helped himself lavishly to my gin.

'But why, James?' I asked.

'Too far away. End of the world,' he explained, waving a dismissive hand. 'Who the hell d'you think is going to come to some remote bloody island in the English Channel to see your set-up? The whole thing's lunatic. I wouldn't come that far to drink your gin. That's a measure of how silly I think your scheme is. Ruination staring you in the face. You might just as well set up on Easter Island.'

This was blunt but not encouraging.

I went to see Jean Delacour at his famous bird collection in Clères. Jean was the most incredible aviculturist and ornithologist, who had travelled extensively, gathering birds, describing new species, writing massive and comprehensive tomes on the ornithology of remote parts of the world. In both world wars his enormously valuable collection of birds had been overrun and destroyed by the Germans. When the last war ended, instead of giving up as most people would have done, Jean started his collection for the third time from scratch at Clères.

As we walked around the wonderful grounds admiring the birds and mammals, Jean gave me a lot of good advice about my scheme, and coming from a man with his vast experience his advice was invaluable. Presently we went down to the edge of the sweeping, velvet lawn where on the edge of the lake tea had been laid.

We sat there, listening to the happy songs of the gibbons on their island in the lake and watched solemn troops of flamingoes, pink as cyclamen buds, crossing the green lawns, accompanied by pheasants and shining jungle-fowl, peacocks negligently trailing their jewel-encrusted tails. Presently, I decided to get France's greatest ornithologist's views on the conservation scene.

'Tell me, Jean,' I said, 'you've been a conservationist now for over sixty years . . .'

'Yes,' he agreed. He was a massive man, with a huge head, not unlike that of Winston Churchill, and with an accent which Maurice Chevalier would have envied.

'Well,' I said, 'what are your views? Do you think there is any hope?'

He brooded for a moment, hands clasped over the end of his walking-stick, his chin on his hands.

'Yes,' he said at last, 'there is hope.'

I was delighted at so unpessimistic a view from such an eminent source.

'If we take up cannibalism,' he added.

I then went to see Sir Peter Scott who was, as always, enthusiastic and helpful. Almost alone among the top ranks of conservationists, Peter believed in captive breeding and it was one of his reasons for setting up his now world-famous Wildfowl and Wetlands Trust. He was most encouraging and constructive about my plans and gave me lots of valuable advice, pointing out pitfalls he had encountered. As he talked he was busy finishing off a large canvas, a painting of sunrise over a marsh with a skein of geese coming in to land. As his brush moved and pecked away at the canvas the painting grew miraculously out of his apparently random daubs, and I remembered a story that had been told to me by a friend who was a painter of pedigree race and shire horses. She had gone to Peter for advice on her first show and he received her affably in his studio, wearing a polychromatic silk dressing gown. As he talked he continued work on the picture he was painting, a flock of ducks coming in to land on a marsh in a sunset. He was in the middle of giving her shrewd and excellent advice when the telephone rang.

'Damn,' said Peter, staring moodily at his canvas. Then he brightened.

'Here,' he said, turning to my friend, 'you're a painter – just fill in all these ducks' beaks with yellow, will you, while I answer the phone.'

Fortunately, Peter did not require such artistic skills from me in return for his help.

I really did not feel my scheme would have the Good Housekeeping seal of approval, as it were, unless I had outlined it to that doyen of the biological scene, Sir Julian Huxley, and received his approbation. He had always been kind and helpful in the past, but this was a somewhat grandiose idea and I was fearful that he would treat it in some damning way. I need not have worried, for he greeted it with the infectious enthusiasm he showed for every new idea, great or small. I relaxed and we had a delightful tea, with talk ranging from the monkey-puzzle forest in Chile to the skin of the giant sloth found in Patagonian caves, from the feeding habits of narwhals to the strange tooth adaptation of a lizard I had captured in Guyana, an adaptation which enabled it to catch, crush and masticate the most enormous snails with great ease. I had sent him a series of photographs depicting the whole process and he had been fascinated.

'Talking of photography, Durrell,' he said at the end of tea, 'have you seen that film young Attenborough brought back from Africa on that lioness . . . you know, Elsa? It was reared by that Adamson woman.'

'No sir,' I said. 'Unfortunately, I missed it.'

He glanced at his watch. 'They're repeating it this afternoon,' he said, 'so we'll catch it, eh?'

So the greatest living English biologist and I perched on upright chairs in front of the television and Huxley switched the set on. In silence we watched Joy Adamson chasing Elsa, Elsa chasing Joy Adamson, Joy Adamson lying on top of Elsa, Elsa on top of

Joy Adamson, Elsa in bed with Joy Adamson, Joy Adamson in bed with Elsa, and so on, interminably. At last the show ended and Huxley leant forward and switched off the set. He mused for a moment. I was silent.

'D'you know what, Durrell?' he asked suddenly.

I wondered what penetrating and lucid commentary on animal behaviour the greatest living English biologist was going to vouchsafe to me. 'What, sir?' I asked, and waited breathlessly for his answer.

'It's the only case of lesbianism I have ever seen between a human being and a lioness,' he said, quite seriously.

After that, I felt that any further conversation would be an anti-climax, so I left.

On 14 March 1959, the Jersey Zoological Park came into being. The first animal inhabitants were an assortment of beasts I had brought back from West Africa and stashed away in my sister's back garden in Bournemouth (that most salubrious of seaside resorts) against the day when they would become founding members of the zoo. They were shipped to Jersey and my sister's neighbours heaved a collective sigh of relief.

Of course, for several months before the animals arrived, Les Augres Manor was a scene of frenzied activity. Carpenters and masons rushing about laying cement, making cages out of everything they could get their hands on. Cages on legs we called them, made out of untreated wood, chain link and chicken wire. Packing crates were wonderfully converted into shelters and every available piece of iron piping or wrought iron from the local junk yard was grist to our mill. We transformed the things people discarded as being of no further use into animal havens and shelters: cages ungainly and ugly but serviceable sprouted everywhere.

Our setting was, of course, idyllic. The beautiful manor house – the oldest fortified manor in Jersey – sat complacently within its granite archways on the edge of a gentle valley, through which meandered a tiny stream which eventually fed a small lake, tree enshrouded. The whole manor was cosseted on all sides by minute fields, each guarded by a hedgerow of trees and bushes, ancient oaks and chestnuts. When – as is reputed – Bonnie Prince Charlie in his bid for the English throne came to take tea on the lawn in front of the manor, most of these magnificent trees must have been mere saplings. One could easily see how, with careful love, attention, pruning and planting, this property could be converted into a park like a ring of greenery with the manor house as the jewel in the setting.

The first big snag soon appeared. It is all very well to borrow £25,000 but this had to be paid back. This meant going on another expedition as soon as possible to get material for another book. So, with the utmost reluctance, I engaged a manager, a friend of some years standing whom I thought I could entrust with the task. This was a mistake. I returned to find that my written instructions and plans had been ignored and the money frittered away. Our ship (our potential Ark, if you like) was an exceedingly frail one and now the hideous shoals and reefs of bankruptcy loomed ahead. It looked as though my plan to create a place to help save animals from extinction was liable to become extinct before it could do any good work. I sacked my manager and took over myself.

The next couple of years were, to say the least, nerveracking. Each morning when I awoke I wondered if that was the day my credit was going to run out and my dream evaporate like dew. The staff were wonderful. Though working on a pittance, they were apprised of the gravity of the situation and all agreed to stay on. This was a great morale booster and gave me the courage (not unaided by tranquillizers) to go out and seduce bank managers into agreeing overdrafts and fruit and vegetable merchants into waiting patiently for their money. Gradually, very slowly, we began to swim instead of sink.

In those early years there were many bizarre happenings and even my mother was subjected to the sort of episode which can occur only if you are unwise enough to live in a zoo. Our two half-grown chimpanzees Chumley and Lulu had discovered, after much research, that interlink wire – if you could find a free end – could be unravelled like an old Fair Isle sweater and almost as quickly. This they proceeded to do to the wire on their cage one afternoon when no one was around. My mother, having just settled herself with a pot of tea in front of the television, heard a peremptory bang on the front door. Puzzled, she went to open it and found Chumley and Lulu on the front stairs. It was obvious from their demeanour that they had come to call, were delighted to find her at home and were in no doubt that she would greet them with the same enthusiasm with which they were greeting her. My mother measured four feet eight inches high and the chimps came up to her waist. Not one to lose her head in a crisis, nothing daunted, she invited the apes in as she would honoured guests, sat them down on a sofa and opened a large box of chocolates and a tin of biscuits. While the chimps were raucously feeding on this manna from heaven, my mother quietly phoned downstairs and reported the whereabouts of the truants. The fact that the apes could have seriously injured her did not occur to her and when I remonstrated with her for letting them into the flat she was puzzled.

'But dear,' she said plaintively, 'they came to *tea*,' and she added thoughtfully, 'and they had jolly sight better manners than some of the *people* you've had up here.'

At one time in the early days we owned an enormous and very beautiful Reticulated python called Pythagoras. Fully twelve feet long and as thick as a rugger blue's thigh, Pythagoras was a force to be reckoned with. He occupied a cage in the then Reptile House, which had been badly designed and which he was rapidly outgrowing. The cage had not been designed by me, I hasten to add, but by the manager I had put in charge in my absence. The front consisted of two large sheets of plate glass which slid over each other, making it extremely difficult to clean if the cage contained a potentially lethal creature like Pythagoras, unless you removed him first. This was a three-man job, two to restrain Pythagoras (who strongly objected) and bundle him into a giant clothes basket, while the third man cleaned out. Though the python was fairly placid as a rule, he strongly disliked being manhandled, and so it was forbidden that any member of staff should try this procedure alone. John Hartley, straight from school, a handsome lad built on the lines of a giraffe, had been with us a year and showed such enthusiasm for the work that we put him in charge of reptiles. One evening his enthusiasm got the better of him. Passing the Reptile House at dusk after the zoo had closed, I heard muffled shouts for help emanating from inside. Investigating, I found John had done the unforgivable. He had tried to clean out Pythagoras alone. The great snake had thrown its coils around him and bound him as immobile as if in a straitjacket. Fortunately, John still had hold of his head, and Pythagoras was hissing like a giant kettle.

This was no time for recriminations. I seized the reptile's tail and began to unwind him. The problem was that as fast as I unwound him from John he threw his coils around me. Soon we were both as inextricably linked as Siamese twins, and we both started to yell for help. It was after hours and I feared that the staff would have gone home. The idea of standing there all night until someone found us in the morning was not a happy one. Fortunately, our cacophonous cries were heard by a member of the mammal staff and with his help Pythagoras was restored to his rightful home. I was, as may be imagined, extremely terse with John. However, being linked together by a python seems to form some sort of bond, for John is now my Personal Assistant.

For the most part we didn't and still don't consider these sorts of episodes as interruptions to our lives, because they are part and parcel of our lives and work. It is only when we take friends or acquaintances around the collection that it is brought home to us that, to the average person, we must lead a very bizarre existence and yet – in spite of thinking us eccentric in the extreme – they are impressed. Today they see our glittering array of reptiles, snakes moving with infinitely more grace than a Balinese dancing girl, tortoises lumbering about like huge animated walnuts. We show them our wonderful group of chocolate-brown gorillas, growling like bears, the leader Jambo like a Sumo wrestler in fur, but much more handsome and gentle as a kitten. Then our shaggy Buddistic Orang utans with their oriental eyes and fur like a hundred tangled pony tails in blond, orange and red. They marvel at our tapestry of birds, cranes as slim and elegant as spears, pheasants wearing plumage of multi-coloured shot silk, flamingoes moving slowly across the green sward like blown rose petals. They fall in love with our tamarins and marmosets, smallest of the monkeys, clad in brown, orange or black fur or a pelt that glistens like pure gold, tiny fragile animals moving like quicksilver through the branches, delicate as birds and trilling and whistling like them. Then in the woods along our lake the lemurs, parti-coloured as dominoes, roaring in chorus so the ground vibrates beneath your feet. Then the babirusa, surely the most beautiful ugly animal in the world, with its great curved tusks and almost hairless body covered with as many folds and wrinkles, nooks and crannies, as a relief map of the moon. The cheetahs, sitting bolt upright in a picture frame of tall grass, the black tear stains on their faces, tear stains – so it is said – because after being created they became haughty and unkind to other animals and so were admonished by God and cried black tears which stained their faces as a reminder of His wrath.

Our friends see all these: animals they know about, others they had never known existed, and they ask how and why we set all this up. We answer that we have over a thousand animals in the collection and ninety per cent of what we have shown them are creatures threatened with extinction and that they come from all over the world. They are threatened primarily by man's activities, and their plight shows what we are doing to the planet. Our *raison d'être* is to provide sanctuary for these creatures, and this is the reason I wanted my own zoo.

Even during the worrisome early years, I decided that we must press ahead with plans for turning the zoo into what it had been created for: a scientific, charitable Trust. However, before the Trust could be created and take over the zoo and run it as the headquarters, there was the zoo debt to consider. Although our gate money was climbing steadily, there was still the wretched £25,000 constantly on the horizon like a black cloud. It was patently obvious that you could not start a Trust in debt for this amount. There was nothing for it if we were to proceed and proceed swiftly: my books were doing well and I took over personally the repayment of the loan so that the Jersey Wildlife Preservation Trust could be born unencumbered by debt.

It was a great day in 1963 when we assembled in the dark depths of the impressive Royal Court in St Helier to hear ourselves incorporated and thus made legal. Lawyers, like black crows in their gowns, flitted through the gloom, their wigs white as mushroom caps in the shade of the forest; they all chatted together in hushed voices, talking that strange lawyers' language which sounds as incomprehensible as Chaucerian English, and when written down is as mysterious as the Dead Sea Scrolls, and sometimes almost as archaic. So finally we emerged blinking into the spring sunlight and went to the nearest hostelry to celebrate the fact that the Jersey Wildlife Preservation Trust was no longer a dream but a reality.

We had certain vital things to do if the Trust were to prosper. We needed a membership, the lifeblood of any organization, and the acquiring of one can be a slow process. In our case, fortunately, the process was accelerated because ever since I had started writing I had kept every letter of appreciation I had ever received. These kind people were now approached and I am delighted to say that a great number of them agreed to join as founder members. (From then on our membership grew until it has reached, at the time of writing, twenty thousand, spread all over the world.)

One of our first tasks after the Trust came into being was a dismal one. Among the animals in the collection was a great number of species which were not endangered in the wild state, animals I had collected when I worked for other zoos, or animals which had been 'wished' on us. They were taking up valuable space and costing money to maintain, when both space and money could be put to better use. So it was essential that these commoner animals were weeded out of the collection and found alternative homes. To dispose of animals, many of which you had hand-reared, many personal friends of long standing, was an unpleasant task but necessary if the Trust were to accomplish what it had been set up to achieve. An additional difficulty was, of course, the paucity of good zoos to which I would dream of sending animals. In the British Isles they could be – with great difficulty and without in any way running out of digits – counted on the fingers of one hand.

So, on this particular morning, Jeremy Mallinson and I walked around the grounds, determined to be ruthless in our choice of the animals we had to dispose of. Jeremy is our Zoological Director and he had joined our ranks just a few weeks after the zoo opened, coming to us for a temporary job: a temporary job that has lasted thirty years. With his Duke of Wellington nose, his buttercup-coloured hair and his bright blue eyes, Jeremy was as devoted to our animals as if he had given birth to each of them personally. His habit of referring to human male and female acquaintances as 'fine specimens' was an indication that his job tended to creep into his everyday life.

Our first stop was at the tapir paddock. These South American beasts are about the size of an elongated Shetland pony, and vaguely resemble a cross between a prehistoric horse and a truncated mini-elephant. Because of their strange, prehensile noses they were called Claudius and Claudette and their baby was Nero. They moved over the paddock towards us, uttering tiny falsetto squeaks of greeting, sounds which are ridiculously minuscule coming from such portly, chocolate-coloured beasts. I remembered, as I scratched Claudius' ear, how I had found him squatting, benign but depressed, in the window of an animal dealer's shop in Buenos Aires. My Spanish not being up to such a situation, I enlisted the help of one of the most beautiful women I have known, Bebita Ferreyra, to help me with the bargaining. She swept into the dingy shop and within seconds, with a combination of charm and fishwife-like shrewdness, she had so captivated the owner that Claudius was purchased for half the price.

Bebita then put on her long white gloves, which she had removed at the outset of bargaining, the better to gesticulate, and left the shop in regal fashion, followed humbly by me leading Claudius on a rope. She hailed a passing taxi, but when the taxi driver discovered that it was Bebita's intention to have Claudius accompany us, he expressed horror.

'Senora, *bichos* are not allowed in taxis,' he said.

Bicho is a useful South American word meaning any sort of wild creature.

Bebita gave him the look Queen Victoria was supposed to give people when she was not amused.

'It is not a *bicho*,' she said coldly, 'it is a tapir.'

'It is a *bicho*,' said the taxi driver, stubbornly, 'a wild and savage *bicho*.'

'It is neither wild, savage, nor a *bicho*,' said Bebita. 'However, if you don't want to earn the thirty pesos for carrying him, I am sure we can find a taxi that will.'

'But the police . . .' said the taxi driver, cupidity struggling with self-preservation.

'You may leave the police to me,' said Bebita, and thus we rescued Claudius.

Now he was a father twice over and as handsome a tapir as I had seen. As I was scratching his ears he suddenly gave a loud sigh and fell on his back as if shot. This was the signal for me to scratch his tummy. As I was doing this, Nero, always on the look-out for food, tried to eat one of his father's ears, which made Claudius leap to his feet snorting with indignation. Jeremy told me he was having difficulty in finding a suitable zoo which had room for Claudius and his family and, though I looked suitably depressed, I was secretly rather pleased.

Next we stopped at the peccary paddock. Here Juanita was the founder mother of the herd of these South American pigs. I had obtained her as a baby in the province of Jujuy in northern Argentina, and the moment I had got the animal collection by train back to Buenos Aires Juanita developed what appeared to be pneumonia. The animals were housed in the grounds of the Natural History Museum while I bedded down in a friend's flat; naturally Juanita had to be bedded down in the flat too, so that she could be nursed. She was in desperate straits and I was sure I was going to lose her. Between the Museum and the flat lay the red-light quarter of Buenos Aires and a street called Venti Cinco de Marzo. Here, between our chores at the Museum and nursing Juanita, we used to drop into a café called Olley's Music Bar for a few fortifying jars of wine. Olley's girls soon found out what my friend David and I were doing and of the sad plight of Juanita. Every evening, with the utmost tenderness, they would enquire after her progress and vie with one another in bringing her small presents (I assume bought with what people would call their 'ill-gotten gains') – a box of chocolates, some figs or avocado pears or perhaps some boiled baby sweetcorn. There was great rejoicing when I told them that Juanita had turned the corner and would live. One girl burst into tears and had to be revived with a large brandy, and Olley himself gave free drinks all round. All I can say is, fallen ladies or not, if I were ill in hospital I would like to be sustained by the genuine love and sympathy of Olley's ladies. I was glad to hear that Jeremy was having difficulty in finding a new home for the peccary herd.

Next we came to the abode of a Palm civet called Potsil. These civets look like small, gingery-brown cats with long ringed tails, the coats covered with blurred darker blotches, and curious protuberant amber-coloured eyes with vertical pupils which give them a faintly reptilian look. I had collected Potsil in West Africa when he was newly born and still blind. As soon as his eyes opened and he got his milk teeth, I realized I was rearing a monster. Potsil lived to eat and would fall upon anything, living or dead, that came within reach. He carried the textbook definition of 'omnivorous' to untold lengths. There was nothing he would not throw himself on to with screams of joy, even if it were some revolting titbit rejected by every other species as being inedible. His greatest ambition in life was to consume a human being – a task he did not feel was beyond his abilities. This made cleaning out his cage a hazardous occupation, for though he looked lethargic he could move like lightning when spurred on by his gastric juices. One of my more impressive scars had been donated by Potsil, so I had no mixed feelings about sending him away. I was passing his abode one morning when I came upon a new member of the staff cleaning out Potsil's cage. Seeing that Potsil looked so catlike, the innocent youth had merely picked the animal up by the scruff of the neck and clasped him to his bosom, so that he could clean out the cage with his other hand. Innocence of this sort sometimes protects as the animal is so taken aback. Foolishly, I decided to help.

'Here, let me hang on to the animal,' I said, 'he knows me.'

I bent forward and grabbed Potsil by the scruff of the neck. My next action was going to be to grab his tail. This method kept one from his mouth and claws. Before I could get a hold, however, a voice from behind me said, 'You *must* be Mr Durrell,' in delighted tones. Distracted by the voice, I gave Potsil his chance. Dangling from my hand like a hanged man from a gibbet, he twisted himself lithely, sank his sickle-sharp retractile claws into my wrist and followed them with his full set of teeth, dentures that would have done credit to a baby Sabre-toothed tiger. I suppose one is always surprised at the amount of blood in one's body, because one does not normally splash it about more than necessary. As soon as Potsil's fangs sank into my wrist, like hot razor blades into butter, I appeared to be losing some three pints of this vital fluid per second. Somehow, I stifled my cry of agony and turned it into 'Good morning', as I turned to face two little old ladies, both wearing pixie hats and wreathed in smiles.

'We are so sorry to interrupt when you are playing with your animals,' said the first pixie, 'but we felt we *must* tell you how much we're enjoying our visit to your zoo.'

'Thank you,' I said, hoarsely.

'All the animals look so happy and well fed,' she went on.

'We try to give them the best of everything,' I said, as Potsil, uttering yarring noises of delight, proceeded to eat his way from my wrist down my hand. I was now losing more blood than any heroine in a Dracula movie, but I managed to hold the animal in such a way that the pixies could not see.

'You play with them all every day?' asked number two with quavering interest.

'No, no, not all of them,' I said.

'Just your favourites, like this one?' suggested the elder pixie.

'Yes,' I said, wondering how much blood you had to lose before fainting.

'How lovely – how they must love it. And love *you* too, of course,' said the younger.

'Oh, yes,' I said, as Potsil's teeth grated on my knuckles, 'they . . . er . . . get very attached to you.'

'Well, we won't keep you, we know you're busy,' said the elder pixie. 'We *have* enjoyed ourselves. Thank you so much.'

As they mercifully moved away, I could hear one say to the other, 'You can see he's a true animal lover, can't you, Edith?' Had they known my feelings about Potsil at that moment, they would unhesitatingly have called in the RSPCA.

I said to Jeremy: 'We can definitely get along without Potsil, though to be fair I suppose we will have to divulge his anthropophagic nature.'

'They are very anxious to have him,' said Jeremy.

'Did you tell them he was a ravening monster, compared to which a Bengal tiger suffering from rabies was a mere kitten?'

'No,' said Jeremy, who had the grace to blush, 'but I told them he was a fine specimen.'

'With your command of euphemism and prevarication we should soon get rid of all the more dangerous creatures,' I said hopefully.

Gradually, hardening our hearts, we continued with the process of elimination but it was a job fraught with difficulties, for not only were my feelings and Jeremy's involved, but those of everyone connected with the zoo. It was bad enough to have to make the decision to part with an animal but, having done so, to discover that the creature had its own fan club among the office or other staff was disastrous. Dictation was taken by tight-lipped secretaries, red-eyed, sniffing into their handkerchiefs, shooting cold glances of hate as if you were a reincarnation of Attila the Hun. Strong maintenance men whom you would have thought would not have a sentimental bone in their bodies gazed at you with loathing, their eyes misty with unshed tears. It was an extremely trying time for all concerned but we managed to get through it without a flurry of resignations.

Another task for the young Trust was to start our filing system. We already kept notes on our charges but these were fairly primitive. What we needed was something much more comprehensive, for I felt that having any large collection of exotic creatures without a proper, detailed filing system was like having a library without a catalogue. This meant cards which recorded where they came from, their age, sex and other details normally vouchsafed on an average passport. But in addition we had to evolve cards which covered all the many day-to-day observations. Within a short space of time, we had amassed a huge fund of information on general behaviour, feeding habits, breeding habits and sicknesses and veterinary treatment. Much of this information had never been recorded before, so we were gradually building up unique archives of the utmost importance. We were – believe it or not, it was the early 1960s – ahead of our time, at least in the United Kingdom.

It was about then that I attended a conference at London Zoo on 'The Role of the Zoo and its Importance'. The best paper given, to my mind, was by Caroline Jarvis, now Countess of Cranbrook. In it she succinctly and clearly set out what zoos should be and what they should do to make themselves better. I was particularly delighted because many of the things she suggested zoos should be doing (but were not) we had already had in operation for several years, and one of the most important of these, of course, was our filing system. It was housed in four massive wooden filing cabinets (we could not afford the luxury of metal ones), a much appreciated gift from a member, and this treasure trove was housed in Jeremy's office.

One night, I was awoken by the sound of feet running across the gravel of the forecourt. Running feet at three in the morning denote a crisis of some sort and in a zoo the possibilities are unimaginable. I was out of bed and halfway down the stairs before I was fully awake. The big room below our flat – in those days the offices, now reception – was full of smoke. I ran through to the corridor which led to Jeremy's office and the smoke and heat grew more intense. It is quite surprising how stupidly one can behave in a crisis. My one thought was that in Jeremy's office there was a baby Colobus monkey we were hand rearing (for we had not the hospital facilities we have now) and all our precious files, both of which had to be saved. I flung open the door of the office and a wall of flame leapt at me, tiger bright, removing in a casual fashion a lot of my hair, my eyebrows and bits of my beard. I staggered back and managed to shut the door. It was obvious that in that inferno the baby Colobus and, so I thought, all our files would be destroyed. We could only wait until the fire brigade arrived, which it did, as usual with the slickness and speed of an eel. Within a short time, they had reduced what looked like the Great Fire of London to the size and fearsomeness of a homely, gently smoking Guy Fawkes bonfire. Eventually, I was allowed to step into the acrid, blackened ruins of the office, the oily water swirling around the floor smelling like the interior of a coal mine.

The poor Colobus was dead, of course, and in the midst of this ugly scene stood our four filing cabinets, charred and black as surviving tree stumps after a forest fire. Gingerly, I pulled open a drawer in one of these crumbling pillars of charcoal. To my utter astonishment the contents, apart from being singed around the edges and a bit damp, were quite undamaged.

'Ah, yes,' said a burly fireman who, with blackened face, stood holding a dribbling hose, 'lucky you had all them papers in there or you'd have lost 'em.'

'What d'you mean?' I asked, puzzled.

'These are wood, see,' he explained, 'thick wood. Taken a while to burn through. If you'd had them in one of these here modern files, the metal would have got red hot and every paper would have been burnt to a cinder. The wood saved 'em, see? Slow burning.'

So it was our antediluvian filing cabinets that had saved our valuable records. Sometimes it does not pay to be too modern.

We were now growing apace and beginning to get organized but we were still, I think, a bit of a puzzle to most of the zoo fraternity. We did not follow the rules. What were we up to? It was ridiculous to think captive breeding would ever be taken seriously by the bulk of the conservative conservation world. At that time, of course, this was true to a certain extent, but there were glimmerings of intelligence in both the zoo and conservation worlds – but they were still two worlds and the glimmers were only glow-worm bright. There is a very old adage which, when people are in some confusion about an idea, instructs you to tell them what you are going to say, say it, and then tell them what you have said. Bearing these useful instructions in mind, we decided to organize and host, with the aid of the Fauna and Flora Preservation Society, the first 'World Conference on the Breeding of Endangered Species in Captivity'. As a conference it was a great success but looking back at it now it seems a bit of a hotch-potch. This was to be expected since captive breeding was a patchwork quilt of endeavour as yet unstitched. But it did, I think, give the conception a morale boost in the right direction. It is wonderful that this conference, first held in Jersey in 1972, has become a regular thing, hosted by different zoos and organizations in different parts of the world, a conference for the gathering and dissemination of information.

It was right at the time of the conference that we had two

incredible pieces of luck. We had two half-grown gorillas, N'pongo and Nandi, who were causing us problems. First, since they had no male, they were becoming very butch and this was worrying if we hoped to breed them. Second, they were rapidly outgrowing their accommodation. Then, miraculously, both problems were solved. Brian Park, a Jersey resident, later to become a Trust Council member and later still our chairman, came forward with the munificent sum of £10,000 after he had seen me on local television bemoaning (as I always was in those days) lack of funds for development. We used this windfall to build spanking new accommodation for our gorilla girls, which was splendid but did little to help with their increasingly entangled sex lives. Then Ernst Lang, the director of Basle Zoo, came to our rescue as, if you like, a sort of zoological marriage guidance counsellor. Ernst had been the first man to breed a gorilla in captivity and to get the mother to rear it instead of taking it away and hand rearing it, which had often been the case in zoos lucky enough to breed these wonderful apes. He had visited us in Jersey and approved of what we were doing, and so he now phoned up and said he would let us have Jambo, first mother-reared gorilla in captivity, a proven breeder himself, to ease the delicate situation between our virgin girls. To have a young adult male gorilla, a proven father, offered to you is as rare as having the key to Fort Knox enclosed with your American Express card.

Now we had this whole problem solved, or so it seemed. There was only one thing to plan: that we should make as much publicity out of our new gorilla quarters and the arrival of Jambo as possible. Who were we going to get to open them? At that time there was a handful of well-known conservationists who seemed to open everything. However, what I wanted was someone outside the conservation movement to show that there were people other than biologists and naturalists who were concerned about the plight of the world's wildlife. But of course it had to be a name for the sake of the publicity. After some thought and with considerable trepidation, I decided on David Niven, a consummate actor whom I had long admired. Whether somebody of his international reputation would feel it worthwhile to come to Jersey to open accommodation for gorillas was a moot point. I phoned my agent for advice and he put me in touch with Niven's son, who worked in London. Would his father, I asked tentatively, take kindly to the idea of acting as best man at a gorilla's wedding?

'I haven't the faintest idea,' was the amused reply, 'but he likes doing mad, unusual things. Why don't you write and ask him?'

So I did, and in due course receive the following telegram: 'Delighted to officiate at gorilla wedding on condition I am at no time left alone with the happy couple. David Niven.'

I met David and his delicious wife at the airport and, though they landed in a howling gale and pouring rain, David was in fine fettle. Over dinner he displayed the urbane wit and scintillating charm for which he was famous and the best thing about it was that it was natural charm and not an act. He told me a number of hilarious and unprintable stories about Errol Flynn, for whom he obviously had a regard bordering on adoration.

'But nevertheless,' said David seriously, 'whatever one says about Flynn, there was one thing you could rely on him for absolutely. In a crisis he would *always* let you down.'

The next morning before ten o'clock when the gates opened I took the Nivens around to meet the animals and they were fascinated. Eventually, we ended up at the Orang utans and I introduced them to Bali, heavily pregnant, the sweetest of apes, beautiful, bulging and benign. She lay in the straw, her beautiful little dark almond shaped eyes regarding us with the placidity of a Buddha, her stomach protuberant in orange fur, her breasts, heavy with milk, obvious in a way that would undoubtedly have won her first prize in a Miss Orang utan contest.

'There you are,' I said to David, 'don't you think she looks like the Orang utan's answer to Lollobrigida?'

At that moment, Bali made a loud and unladylike noise from her nether regions.

'Not only looks like,' David admitted, 'but *smells* like her too.'

After a long and excellent lunch, well irrigated by champagne, David started to show signs of restiveness.

'I say, dear boy,' he said, 'is there anywhere I can change?'

I gazed at the immaculate outfit he was wearing, the height of sartorial elegance.

'What on earth d'you want to change for?' I asked, puzzled.

David frowned at me severely. 'D'you think I'm going to attend this event wearing *this*?' he asked, making a derogatory gesture at his impeccable clothing.

'What's wrong with it?' I asked.

'Not good enough,' said David. 'I have brought with me a suit I had made especially for my son's wedding and I intend to wear that. After all, what's good enough for my son should be good enough for the gorillas, wouldn't you say?'

I agreed, and led him into my bedroom, thoughtfully providing him with another bottle of champagne to help the dressing process. Ten minutes later I looked in to see how he was getting on and found him wandering about the bedroom in nothing but his underpants, sipping champagne and looking extremely distraught.

'What's the matter?' I asked.

'I'm worried,' he said.

'What are you worried about?'

'I'm afraid I'll forget my lines,' said one of Hollywood's most famous actors.

'Forget your *lines*? What lines? All you have to do is to declare the place open and hope that the gorillas will be very happy,' I said soothingly, pouring him more champagne.

'But you don't understand,' he said, plaintively, 'I've made up a *speech*. But I'm terribly afraid I'll forget it.'

'How many films have you made?' I asked.

'I don't know . . . about fifty I suppose. What's that got to do with it?'

'If you're experienced enough to do fifty films,' I pointed out, 'surely you're not going to fluff your lines over the opening of mere gorilla quarters?'

'But that's quite different,' he protested, 'in a film if you do it wrong, you can do it again. But you can't open the gorilla house *twice*, can you? It would look so *unprofessional*.'

With the aid of more champagne, I got him into his suit, an extremely elegant dove-grey tailcoat and trousers of the kind worn by aristocratic gamblers on Mississippi paddle steamers in the eighteen hundreds. Telling him that he looked wonderful (which he did) and assuring him he would remember his speech, I hustled the great David Niven out to the new Gorilla Complex where of course he made a speech of immense charm and humour without fluffing a line. However, when it was over and I got him back into the manor and poured him a drink I saw that his hands were shaking. And this, I reflected, was a man who had won a well-deserved Oscar for one of his performances and who had become famous for displaying the utmost charm and sang-froid in any situation.

By now, the early seventies, our breeding successes with rare animals were excellent, and the list of species in our care had grown considerably. This was mostly the result of my own collecting expeditions, but also of purchasing animals from other zoos or even dealers. At that time the commercial trade in rare animals was not illegal, as it is today, and purchase was often the only way to obtain specimens to set up a breeding group. I felt that a good home at the Jersey Zoo, where the animals would prosper and reproduce, was infinitely preferable to their languishing in dealers' shops or potty little menageries. (Today, of course, we and most other reputable zoos exchange or lend rare animals, with no money changing hands.) We still suffered from that chronic disease, lack of funds, but we were moving forward and our reputation was gradually increasing so that people outside the zoo world were beginning to understand what our motives were and not only to applaud our successes but to be generous in their contributions to our work.

It was at this time, just as I was taking off for my little house in the south of France to earn my living by writing a book, that I learnt that the island was going to be honoured by a visit from Princess Anne. At everyone's insistence, I phoned up the powers-that-be who organize such events and asked innocently if they intended bringing the princess to the manor house to meet the animals. I was only enquiring, I said, because I had intended to take off for France but would, of course, delay my departure if Her Royal Highness intended to grace us with her presence. The powers were shocked. Show the princess the zoo? Never! Her schedule was far too tight. Besides, they had other much more stimulating treats for her to enjoy, like the new sewage works (I think it was) for example. Slightly miffed that we were considered of secondary interest to a sewage works, I reported back and our Council said that this was ridiculous. I must phone up again. So I did and said I hoped they were quite sure, as I was going to France and there I intended to remain until I had finished my book. No, came the reply. The princess' interests lay in sewage rather than the salvation of obscure forms of animal life. So I went to France.

I was just getting into my stride in Chapter Two when I got a frantic phone call from Jersey. The princess had asked to see the zoo. Would I please be present? No, I said, I would not. I had been told she would not visit the zoo. I had come to France and there I intended to remain, writing for my bread and butter. I had, of course, every intention of returning, but I felt piqued at their inefficiency and intended to let them stew in their own juice for a bit. There were more phone calls. Bribery, blackmail, flattery and cajolery had no effect. Finally, when it seemed that everyone was going to commit suicide en masse, I said I would condescend to return. Down in the south of France, I could hear the sigh of relief emanating from Jersey.

I had never been involved in such a visit before. My only contact with royalty had been peripheral, waving a small paper Union Jack on the outskirts of a crowd of some hundred thousand on an occasion in London in my youth. I had no idea of the complexity of it, the intensive searches by detectives of every nook and cranny (I asked if they wanted to search the gorillas, but they refused), everyone with stopwatches timing each step of the way. They had allotted twenty-five minutes for me to show the princess 700 animals spread out over twenty-odd acres and explain the functions of the Trust. I felt it would not do my peace of mind any good to enquire how much time they had allotted to the new sewage works.

It was obvious that the visit would have to be taken at a canter rather than a slow, civilized trot, and so it was essential to try to choose the animals in which the princess would be most interested and, moreover, to have them bunched together. The imminent approach of royalty has an odd effect on one, I discovered. What was I going to say to her? All of a sudden our achievements and our aspirations seemed as interesting as a vicar's sermon. The whole thing seemed a great mistake. I wished I was back in France, but I was stuck with it. Waiting for the car to arrive, I felt like someone going on stage for the first time, hands like windmill sails, feet like Thames barges filled with glue, and a vacancy of mind achieved only by having a thorough lobotomy. The moment she left the car and I bowed over her hand, all my whimsies were washed away. I was taking around a beautiful, elegant, highly intelligent woman who asked unexpected questions, who was interested. I wished the retinue of powers-that-be would go away as they shuffled and twittered nervously behind us and, more fervently, I wished the press would go away as they crouched, clicking like a field of mentally defective crickets in front of us. I think this was the combination that was my undoing, that made me commit the gaffe of all gaffes.

We were approaching a line of cages and in one of them, at that time, we had a magnificent male mandrill, whose name was Frisky. He was – and it is a term you can use only for a mandrill – in full bloom. The bridge of his nose, the nose itself and the lips were scarlet as any anointment by lipstick. On either side of his nose were bright, cornflower blue welts. His face, with these decorations, framed in gingery-green fur and a white beard, looked like some fierce *juju* mask from an ancient tribe, whose culinary activities included gently turning their neighbours into pot roasts. However, if Frisky's front end was impressive, as he grunted and showed his teeth at you, when he swung round he displayed a posterior which almost defied description. Thinly haired in greenish and white hair, he looked as though he had sat down on a newly painted and violently patriotic lavatory seat. The outer rim of his posterior was cornflower blue (as were his genitals) and the inner rim was a virulent sunset scarlet. I had noticed that the women I had taken around before had been more impressed by Frisky's rear elevation than the front and I had worked out a silly routine which I now – idiotically – employed. As we approached the cage, Frisky grunted and then swung around to display his sunset rear.

'Wonderful animal, ma'am,' I said to the princess. 'Wouldn't you like to have a behind like that?'

Behind me, I could hear an insuck of breath and a few despairing squeals, as from dying fieldmice, which emanated from the entourage. I realized, with deep gloom, that I had said the wrong thing. The princess examined Frisky's anatomy closely.

'No,' she said, decisively, 'I don't think I would.'

We walked on.

After she had left, I had several large drinks to steady myself and then faced up to the fact that I had – still sticking to the animal motif – made a sow's ear out of a silk purse. I had intended to ask the princess if she would become our patron, but what chance now? What princess in her right mind would consider this when the leading figure in the organization had asked her if she would not consider exchanging her own adequate anatomy for that of a mandrill? One could not apologize, the deed was done.

Some weeks later, prodded by everybody, I wrote and asked the princess if she would become our patron. To my incredulity and delight she replied that she would. I am not sure how much he had to do with it, but I took Frisky a packet of Smarties – whose virulent colours so closely resembled his own – as a thank-you gift.

2

Trail of the Begging Bowl

I t has always seemed to me to be simplicity itself to raise money for things which are of doubtful help to our planet. Most conservation organizations run around after funds like a starving dog after a bone, and their laudable object is to try to save something from the debris of the world. But should you want money to buy a nuclear submarine, a jolly little pot of nerve gas, an atom bomb or two, the funds are miraculously forthcoming.

We have suffered, as other altruistic organizations do, from financial anaemia, and one of my major tasks has been to whizz about with the frenetic energy of a Japanese waltzing mouse, trying to raise funds. I have occupied myself with this unpleasant task (for I do not like it and don't do it well since I have, unfortunately, little of the con man in my makeup) and I have managed to garner riches from quite unexpected places and from astonishing people.

I have had a complete stranger, a Canadian member of the Trust, give me £100,000 for our new Reptile House, just on the strength of saying (when he complained about our old Reptile House) that if he found me the money I would build the best reptile house in the world.

I have had a letter from a schoolboy containing a postal order for fifty pence. He apologized for the smallness of the amount (it was his week's pocket money) but he hoped it would help.

I had a letter from an old-age pensioner, who enclosed two pounds saying it would have been more but it was difficult to manage on the pension. But she hoped this small contribution would aid our work.

We had a phone call from a lawyer in California, who asked if we were 'Gerald Durrell's Stationary Ark'. We said yes, you could call us that. Whereupon he told us that a Mrs Nubel had died and left us $100,000. I had never met her, nor was she a member of the Trust, so we can only conclude that she had read one of my books about our work in Jersey.

One particularly bad winter remains vividly in my mind. It was when I had just sacked my manager and taken over the zoo. As the Yuletide season approached, it brought scant cheer. It might have been the season of festivity for most people, but not for me. It was a time of no tourists and bad weather – not the most spirited of Jerseymen would venture out and walk around the zoo in drizzle and howling gales. It was a time when the animals' appetites increased, when the food bills soared, when you used three times as much electricity as normal to keep your creatures warm. It was a time when the animals became depressed because they had no humans to look at and wonder at. It was a time when you watched a devoted staff, blue-nosed and shivering, tending their charges in three feet of snow and you wondered if you would be able to pay them their next week's wages. It was, as Shakespeare so lucidly put it, the winter of my discontent, and I put on my coat and made my way down into town to be interviewed by my bank manager.

I have the strong, though possibly paranoid, impression that I spent more time in our bank manager's office then than I did in the zoo. It was fortunate that our particular bank manager, unlike so many of his flint-hearted species, was a charming, kindly and understanding man. If bank managers go to heaven when they die (and concerning this and the ultimate destination of tax inspectors there is some ecclesiastical argument), our bank manager is surely up on a pink cloud with a fully paid up harp and wings, for on that bleak day he saved my life. We went through the procedure of greeting each other with the gigantic and false bonhomie invariably found in dentists' waiting rooms, bank managers' offices and condemned cells. Then we sat down and examined the figures. They were the same figures we had examined ten days previously but our bank manager, with well-simulated surprise, found that they had not changed.

'Um . . . yes,' he said, running his fingers up and down the columns of figures as if searching for some error in addition. 'Yes, it seems as if you are going to be a little short of funds.'

I said nothing. There was nothing to say.

'As I see it,' he said, looking at the ceiling, 'you need some funds to see you through the . . . er . . . really bad part of the winter.'

'Two thousand pounds,' I said.

He flinched. 'You can't, I suppose . . . from any source . . . yes, I see . . . well now, two thousand pounds, yes, a lot of money, and . . . your overdraft with us is now . . . let us see . . . ten thousand pounds; yes, and there is no way in which you can . . .? Er . . . I see.' He thought about it. He drew towards him a small pad on which he inscribed a name, address and telephone number. He tore off the slip and pushed it, as if by accident, across the desk towards me. He got up and paced up and down his office.

'On this island, of course, there are many people who would . . . er . . . help you if they knew your plight,' he said. 'I, of course, as a bank manager, as doctors are, am bound by an oath of secrecy. I am in no way allowed to divulge the name, address, telephone number of any client, nor could I divulge the fact that they have substantial assets. It is unfortunate.'

He paused, and sighed deeply with the heavy burden which this oath of secrecy no doubt confers. Then he straightened up and became more cheerful.

'Come back and see me in a few days when you have got to grips with your problem,' he said, beaming and wringing my hand.

I went back to the zoo. I am very bad at asking people for money, even if they owe it to me, but this piece of paper presented me with a problem for which there were no textbook rules, and for which I was totally unprepared. What does one say, on a cold, wild night, when one phones a complete stranger to ask him for £2000?

'Oh, hello, my name's Durrell and I have a problem' – which might make him think that one of the gorillas was giving birth and that I thought him a veterinary surgeon of high degree. 'I'm from the zoo and I have a proposition which I am sure would interest you' prickles with so many innuendoes and pitfalls that I discarded it immediately it crossed my mind. 'Would you like to make a £2000 contribution to my overdraft?' sounded a mite blunt and smacked of the Mafia. In the end, with damp palms and a voice which kept losing itself in a swamp, I settled for something I considered intriguing but not open to misinterpretation.

'Um, my name's Durrell,' I said to the slow, courteous voice which answered the phone. 'I . . . er . . . at the zoo. I have been given your name because I have a problem on which I would value your advice.'

'Well, certainly,' said Mr X. 'When would you like to see me?'

'Well, would now be convenient?' I asked, wise at least in the ways of catching a bird on the wing, but convinced that he would say no.

'But of course,' he said. 'Do you know your way? I shall expect you in half an hour.'

The drive there, through lashing winds, rain falling with brutality, lightning glaring, had all the trappings of a Hollywood drama. All it lacked was Boris Karloff to open the door. Instead, Mr X opened it himself. Tall, with a large, placid face, intelligent eyes and an air of immense charm and general bonhomie – like a big, secure, elderly, freckled retriever – he commiserated with me for being wet, took my coat and made a gesture at the living room in which a television, in vibrant colours, quivered and shone, but with none of the mystery and imagination-stirring qualities of a Dickensian open fire at Christmas.

'Do come in,' said Mr X. 'My father's in there, watching television.' His father appeared to be eighty-five years old, but he might have been younger. He certainly looked half the age I felt, so maybe my observation was not accurate.

'Could we go somewhere private?' I asked.

'Why, yes,' said Mr X, 'come into my bedroom.'

'Thank you,' I said.

I was ushered into a very small bedroom, which contained an enormous double bed. I had never realized how difficult it was to discuss business of any sort in a small room with a double bed the only place to sit. We both simply sat on the edge of the king-sized bed holding drinks, like a virgin honeymoon couple on their first night.

'Well,' said Mr X, 'what can I do for you?'

I told him.

He listened without question.

'But of course I will help you,' he said, endeavouring to pour me another drink, the action on the expensive bedsprings bringing us closer and closer, as if we were on a trampoline. 'How much do you want?'

Feeling like a dishevelled and singularly unattractive courtesan who had found an easy prey, I croaked out the amount. I remember the cheque book being brought out, smoothed delicately with the assurance of power, the figures meticulously filling in the magic amount and then I was free into the wild winter's night again, with the cheque warming my wallet. The man had been tactful and charming and at no time did he give me the impression that (even on the trampoline of a bed) he was anything but in my debt. To do this under the circumstances required urbane charm of immense degree. In return, I was determined that I would call our first baby Orang utan after him. Three months later, my Mr X suddenly hit the headlines. It seemed that he had, allegedly, swindled a large number of sober Jersey citizens out of their wealth and was, in consequence, forced to spend a short time in one of Her Majesty's less salubrious prisons. I wish I had known him a lot sooner. Not only for his charm, but for his happy banditry. He could have taught me a lot.

During the course of my Robin Hood-like career (taking from the rich to give to conservation) I have travelled widely and met with a lot of adventures, many amusing, many less so, but I never thought that in my efforts to raise funds I should be involved with two countries as dissimilar as can be and yet both of which have in their own ways become inextricably entwined with the history of our work in Jersey. One of these is the most powerful and wealthy in the world – the United States of America – and the other is a poverty-stricken, remote island in the Indian Ocean. It was the latter which first attracted my attention.

Lying off the south-east coast of Africa is an enormous hunk of land 1000 miles long by some 300 miles broad at its widest point, looking like a badly-made omelette. It is called by the euphonious name of Madagascar, and it is the fourth largest island in the world. Biologically speaking, Madagascar is probably one of the most fascinating tracts of land on this planet. The reason for this is that in the dim distant past, when the continents were being forged and crushed into shape, pushed to and fro on the red-hot porridge-like surface of the earth like paper boats on a pond, Madagascar became separated from her parent body, Africa and, like a gigantic ark containing a host of plants and animals, drifted down and to the right to take up its present position. This meant that all its plants and creatures continued to develop in isolation and evolved in totally different ways from their relatives on the mainland.

Nearly all of the living creatures inhabiting this extraordinary island are found nowhere else in the world and what a fantastic range there is. The lemurs, from the huge black and white Indri the size of a four-year-old child, down to the mouse lemurs, the smallest the size of a matchbox; woodlice the size of golfballs; a whole range of hedgehog-like tenrecs, some capable of giving birth to inordinate numbers of young at a time; tortoises as big as footstools, and ones the size of a saucer; an orchid so huge and complex that it can be fertilized only by a moth with a super-long proboscis; a modest pink flower which helps in the treatment of leukaemia, and a host of other biological wonders inhabit this rich and fascinating island.

Unfortunately, Madagascar is a fairly typical example of how we are destroying our world. Once mostly forested, rich in plants and animals, the island is now in a decline. Over-grazing by zebu herds (kept as status symbols instead of mere sources of meat) and wasteful and disastrous slash-and-burn agricultural methods, both undertaken by an ever-rising population, have decimated the Malagasy forests so that ninety per cent of them have now disappeared. This not only means the vanishing of many trees and plants but of the creatures that depend on them. In their place comes erosion, changing, drying and rutting the landscape as age wrinkles and dries the human face. Fly over Madagascar today and you can see this giant island bleeding to death, for without the forest the soil is washed down the rivers and out to sea, great streams of laterite, like blood from slashed veins, coiling their way out into the blue of the Indian Ocean.

Needless to say, the fate of Madagascar is of paramount importance to conservationists because at the present rate of habitat destruction hundreds of unique life forms (many of them possibly of great importance to man) will vanish within the next twenty to fifty years, or maybe sooner. But it is difficult to get a conservation message across when the standard of literacy among the peoples of Madagascar is not high and their economic problems are so severe. It is appalling to realize that the French, when in control of the island, laid down large areas as reserves but did little or nothing to ensure that they were adequately run and policed; moveover, they did nothing to ensure that the native inhabitants were aware of the fascinating and important land which was their heritage. Until recently, the man in the street had only one way of learning of the extraordinary fauna found in his country and that was by looking at the backs of matchboxes on which were depicted in a blurred and highly-coloured form a few species of lemur. It was obvious that, until conservation in Madagascar got under way, it was essential to try to build up breeding groups of all the Malagasy fauna one could procure.

We have always thought it sensible to work with a common

species which is related to an endangered one; in this way you can evolve the best techniques for husbandry and breeding, so that when you acquire the endangered specimens you have already had some experience with a similar creature. Thus, say, the keeping of black bears would be experience which would stand you in good stead if you were going to try to found a colony of the much rarer Spectacled bears. So when we decided that we should seriously consider acquiring various Malagasy fauna to establish breeding colonies, we chose first the three species which were readily available, none of them thought to be in any immediate danger of extinction. They were the Pigmy hedgehog tenrec, the Spiny hedgehog tenrec and the Ring-tailed lemur, one of the most handsome of the lemuroids. The tenrecs are curious, primitive little beasts, moving like tiny, spiky, clockwork toys, pulling the flexible skin of their foreheads down over their noses into a prodigious frown of disapproval should you pick them up, giving birth to gigantic litters of young (the largest litters known among mammals, up to thirty-one in one species). Emerging only at night to feed on insects, raw eggs and meat, the tenrecs kept themselves to themselves and could never, by any stretch of the imagination, be described by that degrading word, 'pet'.

Although we had some setbacks to begin with, we had soon established flourishing colonies. Indeed, in the case of the Spiny hedgehog tenrec, almost too flourishing, for over the years we have bred over 500. One of our male Pigmy hedgehog tenrecs which arrived fully adult beat the longevity record for this species; when he died he must have been at least fourteen and a half years old, an incredible age for such a fragile and delicate creature.

We had success too with our Ring-tailed lemurs. These lovely animals, clad in black-and-white and ash-grey fur with a pinkish wash, have long, elegant black-and-white-striped tails and yellow eyes. They look exactly as though they have wandered out of one of the more bizarre drawings by Aubrey Beardsley. When they walk they do so with a swagger, and their tails are held proudly aloft like banners. Ringtails are great sun-worshippers and the moment there is the most transient gleam of sunshine they seat themselves facing it, hands on the knees or outstretched, heads pointing up with eyes closed in ecstasy, while they drink in the health-giving rays. Our first pair were called by the unimaginative names of Polly and Peter (christened before we received them) and there was no doubt that Polly wore the lemur equivalent of the pants. Poor Peter was thoroughly bossed about, pushed off all the more comfortable branches, driven away from the sunniest spots, forced to relinquish all the better titbits to his harridan of a wife. However, he seemed to thrive on this treatment, so we did not interfere. Polly, of course, was a real prima donna, swaggering about the cage, dozing in the sun with her arm stretched high so that her armpits would get the benefit of the ultraviolet rays, or dancing about elegantly trying to catch those butterflies misguided enough to fly through the wire. In a benign mood Polly would also sing for you, a performance as startling as it was unmusical.

'Come along, Polly, give us a song, you pretty girl,' you would cajole and flatter.

Polly would stretch, preen herself and stare with pensive yellow eyes into the distance as if making up her mind whether you were worthy of her talents. A little more flattery and she would suddenly begin, clasping hold of a branch as a singer clasps her abdomen. She flung back her head, opened her mouth wide and threw herself into her song with all the verve and fire of a lemuroid Maria Callas.

'Ow,' she would sing, 'ar-ow, ar-ow, ar-ow, ar-ow.'

She would pause for applause and then plunge into the second verse.

'Ar-ow, ow, ow, row,' she yowled, 'ar-ow, ar-ow.'

The volume and penetration of the sound more than made up for the slight repetitiousness of the lyrics.

When Peter finally plucked up the courage to seduce Polly we never knew, but he must have caught her in one of her rare weak moments, for she surprised us all by giving birth to a fine male youngster. He was an adorable baby with huge, wistful eyes and a little face, pointed, pixie-like ears and thin arms and legs which made him look in the last stages of emaciation, like a lemur Oliver Twist. To begin with, this enchanting baby rode around on Polly, spreadeagled on her stomach, clutching firmly on to her, his little hands and feet buried in her fur. As he grew older, he grew bolder and started riding on Polly's back, like a diminutive, melancholy-looking jockey on a large steed. Once he had observed and absorbed the world, he became more self-assured and his expression changed from one of melancholy to one of mischievousness. He ventured off Polly's back and explored parts of their domain and would then swiftly return to the safety of his mother's arms when imaginary dangers threatened. He danced and pirouetted daintily, sunbathed like his parents and indulged in the liberty of using their tails as swings. He even learnt to sing with Polly; a shrill, rather quavery accompaniment which did nothing to make the song more tuneful or add anything to the lyrics.

Once we had established the Ringtails, we were lucky enough to secure a group of Mayotte brown lemurs, which come from an island in the Comores off the coast of Madagascar. They were large, attenuated animals with pale eyes and woolly, rather sheep-like fur in various shades of chocolate and cinnamon and black. They settled down very well and within a surprisingly short space of time one of the females gave birth. It was then that we learnt, through bitter experience, some of the psychological problems of the male Mayotte lemur when faced with the joys of parenthood. The baby was no sooner born than it was torn off its mother by the male and killed. This infanticidal attitude of the male was a great shock and we had to work out some means of circumnavigating the male lemurs' evil intentions towards their offspring. In each cage we constructed a maternity den, a cage within a cage as it were. As soon as it became apparent that a birth was imminent, the female was locked in the maternity den. Although separate from the male, she could be seen, smelt and touched through the fine-mesh wire by the male. More important still, he could witness the birth and get used to the idea of the female having a baby. Once the baby was firmly established, the female could rejoin the male and he took the baby's presence as a matter of course.

One morning, I was standing in front of the Mayotte lemur cage with Jeremy and we were admiring the antics of one young couple with their latest youngster.

'At the rate they're breeding, we will soon have to start thinking of some more accommodation,' said Jeremy.

'Yes,' I said, 'and a lot of money it will cost too.'

'I know,' said Jeremy, adding wistfully, 'it would be super to have a whole new range of cages for these lemurs, wouldn't it?'

'Yes, it would,' I agreed.

The baby lemur swung adroitly from his mother's tail to his father's, administered a painful bite and danced out of range of punishment.

'I'm thinking of going to America,' I said.

'America,' said Jeremy, 'you've never been there, have you?'

'No, but I'm thinking of going there and setting up a sort of American branch of the Trust.'

'For raising money?'

'Of course,' I said. 'After all, everyone else seems to go to America to get money. I really don't see why I should be an exception.'

'Yes, well, it should be an interesting trip,' he said, thoughtfully. Neither of us knew just how interesting the trip was to be.

I decided that I did not want to fly, because I felt that flying to and over a country gave you no sense of distance and you miss so much. So I was to go across to New York on the *QE II* and then travel over a great deal of the United States by car and train. That all the Americans I met thought that I was mad goes without saying, but at that stage I knew very few Americans so my resolve to view America from the ground was unshaken. I had arranged to lecture in places as far apart as San Francisco, Chicago and New York, so the tour was going to be a long and arduous one. I decided that I would need somebody to act as my watchdog and protector – my minder, as they are now called; someone who would cope with booking hotels, purchasing train tickets and so on, leaving me free to get as many Trust members as I could, enlisted from my audiences and the people I stayed with. I chose an old friend, Peter Waller, who for some years had been connected with the Royal Covent Garden Opera Company and had, in recent years, helped his friend, Steve Eckart, set up the American School in London. Tall, slim, and handsome, Peter looked no more than forty but was considerably older. He had enormous charm and women – especially elderly women – adored him. I felt that he would be the ideal person to protect me from the overbearing American matrons of the blue-rinse brigade, of which I had heard such frightening stories, for it seemed that my trail might be fraught with terrors unknown to a person used merely to the complications inherent in an expedition to catch wild animals in a jungle. Peter turned out to be a charming, lovable companion, who watched over my welfare carefully, though there were times when his Jeeves-like ministrations fell somewhat short of expectations.

Apart from an elegant series of suits I had built specially for the occasion, I took several hundred copies of our Annual Report (a bulky document) and several thousand leaflets explaining the work of the Trust. Owing to some hold-up at the printers, these were delivered only at the last moment and instead of being incarcerated in stout cardboard boxes were done up in shapeless bundles covered with brown paper and spiders' webs of string. There was no time to repack, so Peter and I arrived at the *QE II* looking as though we had sacked a gypsy encampment and come away with all the more unsavoury spoils. An urbane and aristocratic purser (looking like one of Her Majesty's ambassadors) saw that our gypsy encampment was safely stowed in the bowels of the ship and we were shown our cabins.

It was fortunate for me that some old friends of Peter's were travelling with us – Margot and Godfrey Rockefeller and their two children. Margot, who explained that they were the poor Rockefellers, was a most attractive woman, with a beautiful face framed by prematurely white hair and blue eyes as piercing as a hawk's. She had an impish sense of humour and a great gift as a comedienne, being able to screw up her face and turn her voice into a squeaky falsetto, reminiscent of some of the famous Hollywood puppets. Godfrey, by contrast, was a massive muscular man with a great, round, good-humoured face, perpetually smiling, and humourous sleepy eyes. Their children were an enchanting boy and girl, Parker and Caroline, whom I irritated enormously by calling them Baby Rocks.

The beginning of the voyage went smoothly and Godfrey, who appeared to have an unlimited quantity of Scotch hidden in his cabin, insisted that we all foregather before meals for a few drinks. And then we were struck by bad weather. The next morning both Godfrey and Peter had taken to their bunks. However, I had little time to spare for their woes for I had some of my own. During breakfast I had been approached by the elegant and aristocratic purser to be vouchsafed the unpleasant news that my gypsy encampment of parcels had broken loose during the night and that the baggage-room was now knee-deep in Trust literature. Would I, the purser asked, like to do something about it? The thought of having to repack hundreds of Annual Reports and goodness knows how many thousand leaflets was daunting, but Margot came to my rescue. She rounded up the Baby Rocks and the four of us went down to the baggage-room armed with paper and string supplied by the kindly purser, and surveyed the carnage. To say that we were knee-deep in Trust information was putting it mildly. Grimly, we set to work. It took us all day, but finally the last bundle was wrapped and tied.

'Well, thank God for that,' said Margot, examining her grubby hands. 'What a job.'

'But what a story to dine out on,' I said.

'What story?' asked Margot, suspiciously.

'How I crossed the Atlantic on the *QE II* incarcerated in the hold with three Rockefellers who were tying up my baggage.'

'I shall sue,' warned Margot. 'Anyway, no one would believe you. No one would think that the Rockefellers would be so stupid.'

The night before we arrived, we gathered in Godfrey's cabin to consume several bottles of champagne which he had procured to celebrate our arrival in New York the next day. Under the influence of this delicate liquid, Peter was moved to reminisce about his early days in Vienna in the ballet school.

'The discipline, darlings,' he said, clasping his hands and looking skywards. 'The discipline. You have no idea. So strict, but so rewarding.'

'How d'you mean?' asked Godfrey, lying on the floor of the cabin like a beached whale. 'No booze?'

'No drink at all,' said Peter, horrified. 'Hours and hours and hours at the *barre* until you felt your legs were going to drop off. Simply exhausting.'

'And you did all this without a drink?' asked Godfrey, incredulously.

'Not a drop, dear heart, not a tiny sip.'

'Devotion,' said Godfrey, turning to me, 'absolute devotion. Don't see how you can dance without a drink.'

'What else did you have to do?' asked Margot.

'Well,' said Peter, now well into his fifth glass of champagne, 'they made you dance in a little box thing, I forget the name, to make sure that you only had a small area and if you went beyond it you fell off.'

'Dance in a box?' said Godfrey. 'What sort of a box?'

'Well, sort of flat-topped, rather like that,' said Peter, pointing at the small circular table which was part of the cabin furnishings.

'But that's only the size of a sombrero,' said Godfrey, 'you can't dance on that.'

'Mexicans dance on their hats,' said Margot, thoughtfully refilling everybody's glass.

'But Peter's not a Mexican,' Godfrey pointed out, 'he's Irish.'

'The Irish do clog dances,' I said, to confuse the issue.

'Well, anyway, Irish or not, I don't believe he can dance on that table,' said Godfrey with finality, and took a deep draught of champagne.

We should have been warned. The ship was still shifting uneasily from side to side, but we attributed this to the health-giving properties of the champagne and not to the inclemencies of the weather.

'Of course I can dance on it,' said Peter, annoyed at having his prowess denigrated. 'I'll show you what we used to do.'

He pushed the table into the centre of the cabin and looked at it thoughtfully.

'I'm wearing too many clothes,' he said, and with great dignity stripped down to his underpants.

'That's why ballet dancers get a bad name,' Godfrey said, 'always rushing about exposing themselves.'

'I'm not exposed,' said Peter indignantly, 'am I, Margot?'

'Not yet,' said Margot, philosophically.

Peter climbed on to the table and raised his arms above his head, hands and fingers delicately posed. He rose on his toes and looked at us archly.

'Sing something,' he suggested.

After some thought, Godfrey plunged into a barely recognizable version of the Nutcracker Suite. Peter closed his eyes in ecstasy, twirled round and round, did a few *grand pliés*, and rose to his toes to do another twirl when a seventh wave of some magnitude hit the ship. With a squawk, our poor man's Nijinsky fell off the table in a thrashing of arms and legs. Like a baby bird who has been practising flying on the edge of its nest and has lost its balance and fallen into a frightening alien world, thus it was with Peter. He lay on the floor, his face white, clasping his thigh.

'Ow! Ow! Ow!' he screamed, and the resemblance to a lemur was remarkable. 'Ow! My leg! I've broken my leg!'

That, I thought, is all we need. Due to arrive in New York the next morning and my amanuensis with a broken leg.

We gathered around our fallen hero, pressed champagne to his whitened lips, assured him that he was not in imminent danger of needing the last rites and, what was more important, that his leg was not broken. He had severely wrenched his thigh muscles, but there was no break. However, it was so nasty that he would have to go into hospital for an X-ray and treatment. So when we docked in New York my doughty Jeeves was whisked away in an ambulance and I was left to face the perils of the New World on my own.

Fortunately, my first engagements were in the Big Apple, or in places within easy striking distance of it, so I could get on with these while Peter was languishing in hospital and running up such astronomical bills that I was thankful for our insurance. Medicine in America appears to be such a lucrative racket that I am surprised the Mafia have not taken it over.

It was while I was in New York that the young woman who was acting *in loco parentis* while Peter was out of commission kept telling me about a certain Dr Thomas Lovejoy and how wonderful he was. In her eyes, he was God's gift to everything and it was clear that she was deeply smitten. She said that she was trying to arrange a meeting but Lovejoy was in such demand that he was as hard to catch as a will-o'-the-wisp. Then, one morning, we were outside Macy's after doing some shopping when the young woman uttered a piercing squeak of delight.

'Look,' she cried. 'Look, it's Tom Lovejoy.'

I looked, interested to see this elusive paragon. I saw a slight young man dancing down the sidewalk towards us, tousled dark hair, dark eyes with a humorous glint in them, a handsome face with an endearing grin. I could see at once what made her heart beat faster. I liked him at once and felt that he liked me. Having captured this unicorn so fortuitously, we dragged him off to a nearby hostelry and filled him with beer while I told him what I was trying to do in America. He listened quietly and gave me some excellent advice. I warmed to him, especially when I realized he was one of those rare scientists who took his job seriously but could laugh at himself and others. In fact, it was his outrageous sense of humour that formed a bond between us. In conservation work, if you can't laugh you must weep and with weeping comes despair. Tom promised to see me when I returned from my trip, so that we could discuss how best to set up the Trust in America.

Shortly after this, Peter emerged from hospital and we started on our marathon round of the USA.

America was fantastic and that first trip had many highlights. We had arrived in New York in a heatwave with the air warm and muggy, the like of which I had not experienced outside West Africa. The brown smog lay like cumulus clouds in fish-shaped banks among the skyscrapers, so that their tops stood out pristine in the sunlight like huge cubes of sugar, while portions of them were lost in the surly banks of smog. It was incredible and looked just like a Ray Bradbury Martian city. I fell in love with New York altogether, although I don't like cities. When we drove to Chicago, a place I did not care for – and here I was to lecture to an audience of some 2000 – Peter, to compensate for his ballet gaffe, clucked over me like a mother hen. However, in occupying himself with minutiae, he was apt to overlook the broad canvas. Thus we arrived at the auditorium to discover a packed, expectant house and to find simultaneously that we had left half our film on the Trust at the hotel. I am always a nervous speaker and this did nothing to help me. Worse still was to happen in Chicago. At a cocktail party, kindly given by friends to those members of the audience who might be big givers, I was standing by a sofa on which was sitting a slender, grey-looking man, when I was suddenly approached by a rather fearsome looking woman with bright blue hair, a face like a tomahawk and a voice which could have sliced stone in any quarry.

'Mr Dewroll,' she shrilled, 'my names Avenspark, and that's my husband there.' She made a possessive gesture towards the frail-looking man at my elbow. We bowed to each other. 'Mr Dewroll,' she continued, 'my husband and I have travelled two hundred and fifty miles to be here to hear you lecture tonight.'

'That's very flattering . . .' I began.

'Two hundred and fifty miles,' she said, oblivious of my interruption. 'Two hundred and fifty miles, and my husband a sick man.'

'Really?' I said, turning to Mr Avenspark commiseratingly, 'I'm so sorry.'

'Yes,' she went on in that skull-shattering voice. 'He's got wall-to-wall cancer.'

It was, I discovered, difficult to know what to say to a statement like that. 'I hope you get well soon' was hardly adequate or appropriate. While I was racking my brains, I was fortunately rescued by Peter and whisked away. This, in some degree, made up for the forgotten film.

We got on the train to go from Chicago to San Francisco. We were ushered to our stateroom by a small, ancient, black car attendant with snow-white hair, who looked as though he had just walked off the set of *Gone with the Wind*. To my delight, he talked like it too.

'Dis here is yo' compartment, sah,' he said, 'and yo' gen'leman friend is in dis one right nex' do'. I go fetch yo' baggage right away gen'lemen.'

I knew that partitions between the well-designed little staterooms are removable, so when our porter came back with the bags, I asked him to remove the partition.

'Sho' will, sah,' he said, twisting the knobs that hold the sections, and within a short space of time we had a spacious stateroom with two beds, two basins, two cupboards, two armchairs and two picture windows through which we could admire passing America. It had been a hectic day in Chicago and so I thought we deserved a treat.

'And now,' I said, 'my friend and I would like a bottle of Korbel champagne, please.'

'Sho, sah, sho', I'll fetch it fo' you right away,' he said.

He came back – as the train was sliding out of the station and into the countryside – bearing a nicely chilled bottle of that excellent American champagne in an ice bucket.

'Shall I open it?'

'Yes, please, and have another on ice, just in case,' I said.

'Yes, siree,' he said, busy with the cork.

After I had sipped my approval, he poured the wine, bound the bottle up carefully in a snow-white napkin and returned it to the bucket.

'Will dat be all, sah?' he asked.

'Yes, thank you,' I said.

He paused at the door. 'Forgive me fo' sayin' so, sah,' he said, his face split by a wide, white smile, 'but it sho' is a real pleasure to serve gen'lemen who know how to travel, yes siree.'

As we sat there, sipping the wine, I gave Peter the benefit of my views on train travel.

'This really is one of the best ways of travelling,' I said. 'Who wants to be incarcerated in a sardine can 25,000 feet up with a lot of people you know are going to panic if anything happens? Whereas on a train it is quite different. You are not scrunched up in that prenatal position, you can walk about. You travel at a civilized speed, watching the world go by in great comfort and with great service. Moreover, you are on the ground and you know the pilot is not going to apprise you suddenly of the exhilarating fact that the starboard engine is on fire and not to panic. Train travel, my dear Peter, may be slow but it is safe.'

As I said these words, the train jolted as if it had hit a brick wall. Our champagne glasses did a tap dance and split their contents as they danced, and large pieces of wood and metal started flying past the picture window. The train juddered to a screaming halt.

'Do you think,' said Peter nervously, 'that we've hit something?'

'Nonsense,' I said stoutly, 'trains don't hit things.'

'But this is America,' Peter pointed out.

'True,' I said. 'Let's go and see.'

We joined the other passengers on the track and walked down to where our proud engine was standing dolefully in a shawl of steam. It appeared that a gigantic lorry pulling an equally gigantic trailer had endeavoured to race our train across a level-crossing. There seemed to be no rhyme or reason for this murderous bravado. The lorry got across, but we hit the tail end of the trailer and stove it in. We could judge the force of the collision by the fact that our train's thick steel cow catcher had been buckled as if it were spaghetti. So much so, indeed, that we had to wait three hours while they brought up another engine as a replacement. It was the last time I lectured Peter on the joys of train travel. I think he was so glad to leave the dangers of train travel that when we reached San Francisco we had driven some distance from the station before I discovered that he had left my entire wardrobe hanging up in one of our sleeping-car cupboards.

Finally, after falling in love with San Francisco and hating Los Angeles – a misnomer if ever there was one – I undertook my last speaking engagement at an incredibly exclusive and expensive country club where you could not obtain admittance until you had salted away your first million bucks. It seemed an appropriate place for me to go with my begging bowl. The place was littered with expensive *nouveau riche* Americans, the females with violet hair, necks like vultures and the minute, telltale scars of the last facelift showing through the real or fake tan. They were bejewelled like Christmas trees and tinkled like musical boxes as they walked. I felt that if I could lure one of them behind a bush and strip her of her baubles, it would probably keep the Trust solvent for several years. However, my gentlemanly instincts prevented me from applying this technique for obtaining, easily and rapidly, funds to help animals.

The males who belonged to these females were all, it seemed, six or eight stone overweight, with voices that had a rich timbre through regular gargling with gravel, and purple pouting faces like giant babies suffering from nappy rash. They rode around in electric golf carts with fringes on top, so that there was no risk of their losing weight. Everywhere else in America I had met charming, civilized and overwhelmingly generous and gracious people, so this collection of horrors was, to say the least, offputting. My plea to these moon people was supposed to take place after dinner. Dinner was preceded by two hours of solid and lavish drinking on a scale I have rarely seen equalled. A request for a Scotch resulted in something the size of a small flower vase being thrust into your hand, containing half a pint of spirits, four ice cubes, each capable of sinking the *Titanic*, and a teaspoonful of soda in which three or four errant bubbles were enmeshed.

By the time dinner was served my potential audience was well on the way to inebriation. With dinner, of course, came the appropriate wines with each course and then finally brandy in goblets as large as soup tureens. The woman next to me (who was weighed down by a passable imitation of the Crown Jewels) was, presumably for health reasons, on a mainly liquid diet and merely toyed with her food. The few remarks she addressed to me were, it seemed, couched in one of the more flamboyant and incomprehensible middle-European dialects. I nodded and smiled and said 'Yes', 'Really?' 'Oh' and similar intelligent comments. Then came my great moment. The person who had organized this terrifying ordeal rose to his feet and, fighting gamely against the roar of after-dinner conversation, made a short, totally inaudible speech of introduction and sat down rather unsteadily. This was my cue. I rose. Everyone stopped talking and stared at me owlishly.

I launched into my heartrending plea on behalf of the animals of the world to an audience of quite the most unprepossessing mammals I had ever encountered. As I was plouging on, aware that sibilant conversation was starting up on all sides, I became aware of a curious noise at my elbow. Looking round, I perceived, not altogether to my astonishment, that the woman with the Crown Jewels (doubtless soothed by my mellifluous English accent) had fallen asleep and the weight of her jewels had carried her unconscious head down into her plate which, unfortunately, contained the remains of a lavish strawberry soufflé. Her face nestled in this pink

lavaflow which had enveloped her nostrils. As she breathed ster-
torously, the strawberry soufflé bubbled merrily with a loud
gurgling and popping noise, reminiscent of somebody trying to
suck a complex fruit sundae through a straw.

I was glad not only to sit down but to leave this exclusive club
the following day, richer by a hardly munificent one hundred
dollars.

It was imperative that we caught an early morning plane to
New York, for I had radio, TV and the press lined up, as well as a
lecture. So early that morning, having breakfasted and packed, I
shuffled off to Peter's room to make sure that he had not
overslept. A sad and soulful voice told me to come in. His breakfast tray was on the bed but he was nowhere to be seen.

'Where are you?' I called.

'In here, dearest Gerry, in here,' came his faint voice from the
bathroom.

I poked my head around the door and discovered Peter standing stark naked in the bath, clutching a large towel to his bosom,
with an expression of extreme terror on his face.

'What's the matter?' I asked, puzzled by his demeanour.

'Look,' he croaked, pointing downwards.

I looked. A great stream of blood was pouring down his leg and
forming a gory pool in the bath.

'Good God! What have you done?'

'I don't know,' said Peter, looking as if he were going to cry. 'I
think I cut myself with my signet ring when I was drying myself.'

'Well, come and lie down on the bed and let me have a look at
it,' I said, thinking somewhat uncharitably that Peter *would* wait
until we had a tight schedule before slashing himself with a signet
ring. He lay on the bed and I discovered that the ring had sliced
neatly through a largish vein in that part of his anatomy where it
was impossible to put a tourniquet without turning my friend and
helper into a boy soprano. Wondering how on earth to staunch the
blood, my desperate eyes fell on the breakfast tray. Reposing on it
was a saltcellar: quickly, I mopped the blood off the severed vein

and poured the contents of the capacious saltcellar on to it. The results were not quite what I had expected. All his ballet training came to the fore. Peter left the bed with a leap Nijinsky could not have emulated, scattering blood and salt in all directions, uttering long yodelling cries indicative of extreme agony. I had to pursue him several times around the room before I could get him to lie down on the bed again, but only when he had extracted a promise that I would not use any more salt. Of course, all the activity had excited the vein which was now pumping out blood like a fountain. It was obvious that something drastic must be done if Peter were not to bleed to death and we were going to catch our plane on time.

I phoned down to the desk. yes, surely they had a first-aid kit. What was the trouble? My friend, I said, had cut himself, hoping they would believe it was a shaving cut. Could I come down and collect the kit? I surely could. Exhorting Peter to lie still, I sped downstairs. I reached the desk simultaneously with a giggling, happy flock of teenage American girls, who surrounded me on all sides.

'I was so sorry to learn about your friend,' said the clerk, putting the box of medicaments on the counter. 'Where has he cut himself?'

'I . . . um . . . that is to say . . . only a scratch, bleeds a lot, you know,' I said.

The American girls looked at me with interest, their attention focused by my accent. The clerk opened the box and rummaged in it and held up a roll of sticking plaster.

'How about this, sir?' he asked helpfully.

With all the innocent young maidens clustered round me I could hardly explain that sticking plaster would not adhere to that part of the anatomy I wished to practise my nursing skills upon.

'Perhaps if I could just take the whole box,' I said, grabbing it. 'So much easier, you know.'

'Why, surely, sir,' said the puzzled clerk, 'but this is medicated.'

'I'm sure it's excellent,' I said, clasping the box to my chest and backing through the bevy of young females, 'but I'd just like to have a look . . . thank you so much . . . bring it back.'

I made my escape into the lift. Once in the safety of Peter's room, I examined the box, which was large and remarkably well stocked with every remedy known to man except the means of staunching blood. However, delving into this capacious cornucopia, I came across a large aerosol bottle with the label 'Newskin' on it. I gave it an experimental squirt and it ejaculated a fine filament-like, cobweb thing that hardened like crystallized sugar.

'Ah, just the stuff,' I said, to give Peter courage and at the same time wondering what it was. I staunched the blood for a moment with my finger and then took aim with the aerosol and let fly. I think I possibly pressed too hard. At any rate, the aerosol ejected a great cloud of spider's web and in a second Peter's entire external genitalia resembled one of the more flamboyant and interesting South American birds' nests. It certainly stopped the bleeding but I wondered uneasily if the stuff would contract as it dried. However, after a few moments it did not appear to be having any ill-effects so I bundled Peter unceremoniously into his clothes and we fled to the airport, catching our plane with about half a minute to spare.

On our return to New York I had further talks with Tom Lovejoy and we worked out the formula which allowed Wildlife Preservation Trust International to come into being as the American fund-raising arm of the Trust. To be more accurate, I suppose, I said what it was that the Jersey Trust needed and Tom hammered out the master plan of how to obtain it.

There are times when the trail of the begging bowl is a hard one, but in this case it was more than made up for by the wonderful and generous people I met in America and who, moreover, rallied round when I launched WPTI and became our first Board of Directors. Over the years we have had reason to be more than grateful to our American friends, for most of our big gifts and grants have come from across the Atlantic, and without this magnificent help our progress would have been slow indeed. However,

I feel I must point out that among American desirable exports there are other things besides dollars and this is where the lemurs of Madagascar re-enter the story, this time in the unlikely guise of matchmakers. //

Duke University in North Carolina was justly famous for having the largest collection of lemurs outside Madagascar, and their breeding successes and the studies they were carrying out were excellent. So it came as something of a shock to receive a letter from Professor François Bourlière (one of France's foremost primatologists), who is on our Scientific Advisory Committee, to inform me that he had heard the great lemur collection at Duke was to be disbanded owing to lack of funds. Did I think the Trust could do something? There was, of course, nothing we could do to help financially, but if the awful news that the collection was going to be disbanded was true we could, I felt, offer homes to one or more species. I was on the point of taking my pitcher once more to the dollar well, so I phoned our by then established American Board and said that I would like to visit Duke University before starting my new banditry in their country, and they readily agreed.

It was arranged that I would fly to Durham and be met there by the long-suffering Margot Rockefeller, whose Baby Rock Caroline was an undergraduate at Duke. Margot, ebullient as always, met me, and as we drove to the university I briefed her about the importance of Duke's primate collection.

'If it's so damned important, why doesn't the university support it properly?' she asked, logically enough.

'I haven't the faintest idea. I can only conclude that as usual the alumni are more interested in supporting the college football team than what they would think of as a bunch of smelly lemurs.'

'Well, if the collection's as important as that, I think it's a disgrace,' said Margot belligerently.

When we arrived, we found that they had laid out the red carpet and we were taken round in a gaggle of professors who explained things to us. For the next three hours I was in my element, peering at cage after cage of beautiful animals, Red ruffed lemurs gaudy as banners, Ringtails sitting in rows like decorative friezes, Sifakas with their pale, silvery fur and black velvet faces with huge golden eyes, clutching their perches and looking exactly like a Victorian child's toys, monkeys on a stick. There were Mongoose lemurs, clad in fur in a variety of shades of chocolate with their pale eyes making them look strangely predatory, and Mouse lemurs leaping like thistledown around their cages, their walnut-sized heads seeming to consist entirely of huge topaz eyes and delicate, petal-like ears. We had lunch and the conversation was entirely confined to lemurs. Before the weight of this scientific avalanche, I could see poor Margot beginning to fade. I myself, suffering from jetlag, was making fairly heavy weather of it as well. After lunch, we had another two-hour session with the lemurs and then Margot and I staggered back to our motel carrying with us the knowledge that the professors had, out of the kindness of their hearts, laid on a dinner party for us that evening. 'I don't think I can stand it,' said Margot plaintively. 'I wouldn't mind, but I don't understand half of what they say. Do they always have to use words of ten syllables?'

'Yes,' I said sorrowfully. 'It shows you're an academic and not rubbing shoulders with a lot of the uneducated hoi-polloi like you and me.'

'Well, I don't know how I'm going to face this party tonight,' said Margot.

'You needn't come,' I pointed out. 'The party is really for me. I've got to go but you can pretend you've got a fallen arch or something.'

'No, honey, I've stuck it with you so far, I'll see you through tonight,' said Margot in martyr-like tones.

'Come to my room beforehand and I'll give you a nice drink to get you in the party mood,' I said.

Later, with the aid of a bottle of Scotch, we tried to get into the party spirit, so by the time we arrived at our host's house we were flushed and full of false bonhomie. Fortunately, everyone was on their third drink (of the size you see poured only in America) and so our appearance passed unnoticed. All the professors had brought their wives and they talked polysyllabically as well. There seemed to be little hope for Margot and me, and I saw a stricken look on her face. I, too, was gazing round the room desperately, searching for a nook or a cranny to secrete myself in, when my glance fell upon a young woman who was sitting on what used to be called in my day a pouffe, nursing her drink and looking remarkably attractive. I glanced at her hands which were ringless, I glanced around to see whether any muscular young man was exuding a proprietary air and there was none. One of the delightful things about America is that you can introduce yourself to complete strangers without having them faint with horror. So I drifted across to the girl.

'Hullo,' I said, 'I'm Gerry Durrell.'

'I know,' she said. 'I'm Lee McGeorge.'

'What do you do?' I asked, hoping she wasn't going to tell me that she was engaged to one of the professors and that the engagement ring had just gone to be cleaned.

'I'm a student,' she said.

'A student of what?' I asked, hoping she would not say psychology, nuclear physics or historical drama of the late 1600s.

'I'm studying animal communication,' she said, 'at least that's what I'm doing my PhD in.'

I gazed at her in stupefaction. If she had told me that her father was a full-blooded Indian chief and her mother a Martian, I could not have been more astonished. Animal communication in all its forms happened to be a subject in which I was deeply interested.

'Animal communication?' I asked stupidly, 'you mean the way animals communicate with each other, whistles, grunts, honks and so forth?'

'Well, roughly speaking, yes,' she said. 'I did two years in the field in Madagascar, studying the noises of forest animals.'

I gazed at her. That she was undeniably attractive was one

thing, but to be attractive *and* to be studying animal communication lifted her almost into the realms of being a goddess.

'Don't go away,' I said, rising to my feet. 'I'll replenish our drinks and you can tell me all about Madagascar. I've never been there.' So for the next two hours we talked about Madagascar and argued ferociously about animal communication. We may not see eye to eye on everything, I thought, but at any rate we are having no difficulty in communicating with each other, mammal to mammal, as it were.

Then, at ten o'clock, our host rose and said he thought we ought to go to dinner. I had thought we were having dinner where we were, but apparently we had to go to some restaurant. It transpired that Lee was the only one to know the way to this watering hole, so she was detailed to lead us in her car.

'Good, I'll come with you,' I said, firmly, 'then we can go on talking.'

Her car was tiny and for some inexplicable reason full of dead leaves and dogs' hairs. We set off followed by a sort of funeral cortège of professors and their wives, all in a highly convivial mood, bearing Margot Rockefeller in their midst. Lee and I continued our discussion and so absorbed were we that it was some considerable time before we became aware that Lee had taken a wrong turning and was now driving round and round in circles, followed trustingly by the cream of academia. After several abortive attempts, we found the right road and arrived, to the restaurant's frigid disapproval, an hour and a half late. Over dinner, Lee and I continued talking and at about two in the morning she drove me back to the motel.

Next morning I awoke and discovered, not surprisingly, that the slightest movement of my head spelt agony. Lying quite still, I thought about Lee. Had it, I wondered, been an alcoholic haze that made me think her so intelligent? Beautiful, yes, but intelligent? I put in a call to Dr Alison Jolly, the doyenne of Madagascar studies and the winsome ways of lemurs.

'Tell me, Alison, do you know a girl called Lee McGeorge?'

'Why, yes,' she said, 'Duke University.'

'Well, what do you think of her?' I asked, and waited with bated breath.

'Well, she's quite one of the brightest students in the animal behaviour field that I've come across for many a year,' said Alison.

My next problem was not so easily solved. How did one try to attract a young, pretty girl when one is portly, grey and old enough to be her father? To one who has collected mammals successfully in all continents, the problem of this capture seemed, to say the least, insoluble. Then I suddenly remembered the one unique attribute I had: a zoo. I decided that I must get her over to Jersey to see my lonesome asset. But how could I do it without arousing the darkest suspicions in her bosom? This complication occupied my mind for the next few days; then I was struck with a brilliant idea. So I phoned her up.

'Hello, is this Lee McGeorge?' I asked.

'Yes,' she said.

'This is Gerry Durrell,' I said.

'I know,' she said.

'How did you know?' I asked her, taken aback.

'You're the only person I know who would phone me up with an English accent,' she said.

'Oh,' I said, struck by the logic of this. 'Well, anyway, I phoned you because I've got two bits of good news. The first is that I have got a grant that will enable us to build the hospital that we need.'

'Wonderful,' she said. 'That's great.'

I took a deep breath. 'And the second piece of news is that an old woman, a member of the Trust, has died and very generously left us some money in her will. Now normally when people leave money to the Trust they specify what it is to be used for, but in this case she has left it to me personally to use as I think fit.'

'I see,' said Lee, 'so what are you going to do with it.'

'Well, you will remember,' I said, 'that I was anxious to set up a behavioural study and a sound recording unit?' This at least was true.

74

'So you are going to use it for that. What a great idea,' she said, enthusiastically.

'Well, not quite,' I said. 'It's only a small amount of money, not enough to build anything, but enough to do the preliminary research on its viability. So I was wondering . . . if we should use it . . . to bring you over to Jersey to give me advice on setting the thing up. How does that strike you?'

'It sounds a super idea,' she said, slowly, 'but are you sure you want *me* to advise you?'

'Definitely,' I said, firmly. 'With your experience no one could be better.'

'Well, I'd certainly love to do it, but I couldn't come over until the end of the semester.'

So she came over, armed with a massive tape recorder, and spent six weeks in Jersey. As I expected she would, she adored the zoo and the whole concept of the Trust. At the end of six weeks, with some trepidation, I asked her to marry me and, somewhat to my astonishment, she agreed.

Now I am a modest man by nature, but I have achieved one irrefutably unique thing in my life, of which I am extraordinarily proud. I am the only man in history who has been married for his zoo.

3

Complicated Conservation

If written down, the initial stage of using captive breeding as a conservation tool looks completely straightforward. You choose the animal that needs help and you set up a breeding colony. However, things are not quite as easy as that. The saga of the Pigmy hog is a good example. The problems we encountered while trying to help this diminutive member of the pig family taught us many lessons. It taught us that field trips are essential as our knowledge of most species in the wild is negligible. It taught us that in many cases in different parts of the world government inertia or inter-departmental bickering can prove fatal to wildlife conservation. And in this particular instance, it taught us that some animals may not be as endangered as we think, for, at the time we started taking an interest in this little animal, it was thought to be extinct.

The Pigmy hog, smallest of the Suidae, was first described from Assam in Northern India in 1847 by B. H. Hodgson. To begin with, no one was quite sure if this tiny pig was a full species or merely the young of the common Indian Wild Boar. However, it was soon proved to be a new species and was christened *Sus salvanius*. A few museum specimens were obtained and then, as mysteriously as it had appeared, it vanished. This was thought to be because of human encroachment into its habitat, the giant Elephant grass or 'thatch', as it is called, which was (and still is) burnt and then ploughed up for farmland. So it seemed the Pigmy hog had made a brief appearance on the scientific stage only to join the Dodo in oblivion.

However, in an area so vast and so little visited by scientists, it was quite possible that such a small, shy creature could have escaped detection and that it might still be lurking in undisturbed patches of thatch. Making a mental note that I should, one day, go on the trail of the vanishing hog, I thought no more about it until one Captain Tessier-Yandell entered my life. Not only did he enter my life but he was accompanied by an otter, which to me is infinitely preferable to a pallid visiting card. What the captain wanted from me was a temporary billet for his otter. He was shortly to retire from Assam and come to live in Jersey, when his pet would join him. I did not really *want* an otter, beguiling things though they are, but this was such an enchanting creature it soon wormed its way into my heart. While we sat in my office with the otter rippling around the floor in that wonderful, apparently boneless way they have, Tessier-Yandell said he was shortly returning to Assam and would be happy to search for any other specimens I might require.

'Pigmy hog,' I said instantly.

He looked blank, as well he might. 'What's a Pigmy hog?' he asked, hesitantly.

'Smallest of the pig family, thought to be extinct, but I'll bet it's not. Charming little animal,' I said enthusiastically. I had never actually met a Pigmy hog, but I have a deep, warm regard for all members of the pig family. So, simply because it was a pig and a pigmy pig at that, I felt it must be charming. I got the only picture I had of the beast and we pored over it. They stand about fourteen inches at the shoulder and are roughly the size of an overweight wire-haired terrier. They are covered with grey and black bristly fur that looks like spikes and have tiny but serviceable tusks. At first glance they do look very like a baby Wild boar, but closer inspection shows a totally different configuration of the head. Even I, ardent pig lover that I am, must confess that an adult Pigmy hog could not be described as beautiful by even the most dedicated of porcine admirers.

Tessier-Yandell, to my delight, seemed fascinated by the

whole idea. 'I shall certainly keep an eye out,' he said, 'and I shall ask the local people, and see what happens.'

Over the years, I have had dozens of people make the same promises and very few have come to fruition. But in Tessier-Yandell's case he was as good as his word. Within an astonishingly short space of time he wrote to inform me of the exciting news that Pigmy hog *did* exist – albeit in small numbers – and the local people knew of it and were going to try to catch some. He himself unfortunately could not oversee this as he was leaving, but he had passed the whole matter over to the Assam Valley Wildlife Scheme, who were already involved in keeping and breeding the extremely rare White-winged wood duck. Then, after a short while, I got the fantastic news that Pigmy hog had been captured and no less than three pairs had been established on a tea estate in Attareekhat. Four of these precious pigs could become ours if we could accomplish two things: first, get the Indian government to agree to their export and, second, get the British Ministry of Agriculture to agree to their importation into Jersey, for Jersey importations are governed by the same laws as those affecting the United Kingdom.

The first problem was solved by writing to Sir Peter Scott, who was not only on our Scientific Advisory Board but was also Chairman of the Species Survival Commission of the International Union for the Conservation of Nature and Natural Resources (IUCN). He at once wrote to Prime Minister Indira Gandhi, always a champion of conservation, and she agreed immediately that we should be allowed to export two or three pairs of Pigmy hog to Jersey. That seemed to finalize that part of the operation, or so we thought. The Ministry of Agriculture was, if I may say so, a different kettle of fish. Most of the Ministry vets go white and faint if you suggest bringing cattle or sheep or goats or any cloven-hoofed animals into the United Kingdom for fear that they might sully pure British stock with such filthy foreign diseases as anthrax, rinderpest, blue tongue, foot and mouth, and other such obnoxious and contagious sickness. Pigs, in particular, give the best a collective nervous breakdown lest they infect the noble British pig with swine fever, or Aujesky's disease.

After a prolonged correspondence which started coolly but ended on a more human note, they said, reluctantly, that we could bring the pigs to a European zoo and breed them there. Then, if it could be proved beyond the shadow of a doubt that none of these nasty ailments had been found in the area in the past six months, we could import the young to Jersey. This sounded like the answer to our difficulties, but not quite. Europe also has its quarantine laws and our problem now was to find a suitable zoo which was allowed to import wild pig, had suitable accommodation and *wanted* Pigmy hog. The whole thing had now become so complicated I was sorry I had ever heard of the animal. Then, just as we were beginning to despair, Zurich Zoo came to our rescue. They would have the pigs, try to breed them and, if successful, the progeny (or some of them) would be sent to us. Flushed with triumph (this had taken six months to achieve), I decided that Jeremy should fly out to Assam immediately and ship the little animals back in his care. Of course, when Jeremy got out to Attareekhat, he ran straight into that bane of the conservationist's life: politics.

For some years (and it is still going on) Assam had been trying to break away from India and become self-governing. So when Jeremy arrived, full of enthusiasm, he found that relations between India and Assam were, to say the least, extremely prickly. Thus when he gaily said that he had come to collect three pairs of Pigmy hog and showed his authorization from Madame Gandhi, it cut no ice with the locals at all. He began to feel about as welcome as an undertaker at a wedding. In vain did he appeal to the local authorities. The Chief Conservator of Forests, the man with the power, simply said he did not have enough pigs to spare. To spare from what? one wondered. He was doing nothing with the ones in his care and was patently doing nothing to protect the habitat of those pigs left in the wild which, after all, was his job. Faced with political antagonism and bureaucracy, Jeremy nearly went mad. Cables to and from New Delhi had no effect. The Conservator of

Forests was adamant. Jeremy felt he could do no more and was just about to concede defeat when the Conservator decided that he had played the political game to its limit and it would be perhaps unwise to continue to be obdurate. He had made his point to central government. Magnanimously, he said that Jeremy could have a pair of Pigmy hogs. Jeremy was now in a quandary. When the Trust was formed we had decided, after much careful discussion and consultation with scientists that the minimum number of specimens of any species to found a breeding colony with a wide enough gene base would be three pairs, and this should always be aimed at except in exceptional circumstances where, say, the wild population itself consisted of only eight or ten specimens. After careful thought, Jeremy decided, quite rightly, that after all the time, energy and money we had spent on the project we had better make the best of a bad job. So he bundled the two pigs in crates and made his escape before the Conservator could change his mind.

The little creatures arrived safely in Zurich and settled very well into their quarantine quarters. Their adaptation to captivity and a strange diet was a great success and we had high hopes. To our delight, the female gave birth to a litter of five piglets. Unfortunately, these proved to be four males and one female. It now became obvious that our strategy to obtain three pairs of a species was a wise one, for we now had a preponderance of males. The original two parents suddenly died, leaving us with the sexually uneven group of babies. However, they continued to prosper and grow. Then, when they had reached maturity, Fate dealt us a nasty underhand blow. Our one remaining female piglet died in childbirth. This left us with four lovely young males all on their own. By this time relations between Assam and India had reached an extremely unpleasant stage and so our chances of going out in search of more females was nil. Once again, as had so often happened in the past, politics was impeding the progress of conservation. In desperation we obtained sperm from our pigs and artificially inseminated some domestic Gottingen miniature pigs, hoping for female offspring which could then be bred back to their Pigmy hog uncles, and eventually produce descendants which would be very close, genetically speaking, to a 'true' Pigmy hog. Out attempts failed because the Gottingen pigs didn't even conceive.

So this was the dismal tale of the Pigmy hog, thought to be extinct and rediscovered, of a rescue attempt that failed, and now the little animal has sunk back once more into obscurity. When we first got involved in this saga, 40–50 per cent of the thatch area, the only known habitat for the Pigmy hog, was being burnt each year. Latest reports say they are burning 100 per cent per annum. As if that weren't enough, the last stronghold of the Pigmy hog, the Manas Tiger Reserve, which is home not just to the hog but to the terribly threatened Great one-horned rhinoceros and the Wild buffalo, has been invaded by armed tribal dissidents. These people have killed forest guards, set fires and shot rhino. Although the Indian army has stepped in and the situation is said to be now under control, it seems as if the Pigmy hog, having escaped extinction by the skin of its teeth, is now definitely heading on the downward path to join the Dodo, the Quagga, the Passenger pigeon and a host of other creatures which tried and failed to live with the most monstrous predator of them all, *Homo sapiens*, a misnomer if ever there was one.

Extracting facts from the other animals being a difficult task, you would think that communication with one's own species would be a fairly straightforward business, for even a language barrier can be overcome. I have learnt to my cost that this is not so, and that the extracting of information from your own species can be as difficult as unravelling the sex life of some obscure deep-sea fish. This was brought home to me when we got the White eared pheasants.

These graceful and beautiful birds inhabit the highlands of China and Tibet and probably in the wild state (like most game birds, the guans and curassows of South America, the guineafowl of Africa, and so on) are getting increasingly rare under pressure from hunting and habitat destruction. The last White eared pheasants to be exported from China had been in 1936 and the captive population that was now left consisted of some eighteen birds, all either too old or incapable of breeding. So when we got the chance of obtaining some new birds from China to establish a fresh and viable captive colony, we leapt at it. We obtained two pairs and I have written elsewhere of our trials and tribulations in trying to breed them. Finally, however, against great odds, we succeeded, and it was a red-letter day for us when Shep Mallett, our then Curator of Birds, and I stood gazing fondly at no less than thirteen delicate and fluffy babies clad in pale fawn down marked with chocolate blotches, who peeped and trotted their way around their bantam foster-mother like so many of those wind-up clockwork toys you can get from street vendors. We had, of course, started a complex file on these valuable babies, but certain vital information was lacking regarding the current status of the species in captivity and in the wild. We learnt from a Dutch dealer, from whom we had bought the birds, that he had obtained them from Peking Zoo, so what more natural than that I should write to the director of the zoo in pursuit of the information I needed?

I wrote a glowing letter about our excitement at receiving the pheasants, gave him details of the Trust's work and asked for his assistance. Enclosed with the letter I sent copies of our Annual Report, our guide book and photographs of the baby pheasants and their parents in the aviaries. Days passed and stretched into weeks. I felt that, with the upheaval of the cultural revolution, my letter might have gone astray. So I sent off a copy of my original screed (plus more photographs etc.) with a cover letter saying that I felt sure my original letter had gone astray. Weeks passed and nothing happened. I wrote a third and eventually a fourth time, with the same result. After some thought I decided on a new plan of campaign. I wrote to the Chinese Ambassador in London, enclosing copies of my various letters to the Peking Zoo and asking for his advice and help. Nothing happened. I wrote again

saying that I felt sure my letter had been mislaid by the foul deca-
dent British postal authorities and I included copies of everything
I had ever written about the White eared pheasant. There was no
reply. It was as though I had never set pen to paper. By now, feeling more than slightly wrathful (after all, I was not asking for
details of one of their atomic sites), I sat down and wrote to the
Chargé d'Affaires at the Embassy in London explaining the situation and enclosing copy letters. This pile of papers had by now
swollen to the proportions of a manuscript by Tolstoy and had cost
me a small fortune in postage. Silence reigned. I wrote again, twice.
Silence. Feeling desperate, I carefully copied out my mass of unan-
swered correspondence to the Chinese and sent the lot to our
Ambassador in Peking, apologizing for the trouble I was causing
and begging his help in breaking through this silent curtain. He
wrote back courteously to say that he had forwarded my correspondence to the director of Peking Zoo and that was really all he
could do. He did hope that I would receive a satisfactory reply.
Not altogether to my surprise, I got no reply at all and now, nearly
thirty years later, this is where the matter rests. Trying to cope
with a one-sided correspondence takes on all the futile aspects of
sending letters up the chimney to Father Christmas.

The Latin Americans also make it a point of honour not to
answer letters, or at least they did when we first got involved with
the Volcano rabbit.

The Volcano rabbit is peculiar enough to have a genus of its
own and to live only on the slopes of volcanoes outside Mexico
City, the ones with unpronounceable names, Popocatepetl and
Ixtacihuatl. It is diminutive, being roughly the size of a baby European wild rabbit, but with smaller, neater ears, tucked close to the
head, a somewhat more rounded profile and a very alert stance all
its own.

I remember the first day we drove up the flanks of
Popocatepetl, looking for the rabbit, to where the road ended and
the great snow drifts began, crisp as any Victorian pie crust. We
had seen nothing except scattered pine forest and a sea of Zacaton

grass, great golden tresses on hummocks, looking like the wigs of a hundred thousand courtesans on wig stands. Then, as we drove back down through the clear air towards the valley of smog in which Mexico City crouches, a city with increasing halitosis, we suddenly heard a strange noise, like a cross between a chirrup and a bark and there, on top of an elegant wig of Zacaton, sat a Volcano rabbit, compact, erect, watching us with circumspection. I looked at this delightful little creature, looking as if newly washed and brushed, watching us with his small bright eyes, sitting in his king- dom of Zacaton tussocks in the clean, crisp, thin air on top of the volcano. Then I looked down into the valley where the huge sprawling city lay invisible under its thick haze of smog. I thought that the Volcano rabbit knew how to fit into his environment without despoiling it, whereas man, wherever he goes, seems to foul his nest, ruining it both for himself and for the other creatures which try to exist with him.

Even in the sixties, you could see that cattle and crops were spreading out from Mexico City and creeping up the slopes of the volcanoes, threatening the rabbit's habitat, and so I thought it was time to do something about it. Over a period of about two years, I wrote eleven letters to the department concerned in Mexico City and received no reply. In a fit of annoyance, I sent some of thee registered so that no one could say they had not been received. When, finally, I got so exasperated that I decided the only thing to do was to go to Mexico myself, I made several appointments to see the person I had been 'corresponding' with, most of which were cancelled at the last minute without excuse. When, finally, I got in to see him, he blandly denied all knowledge of my correspondence though I showed him my file of copy letters. Then, of course, to salve his Latin American *amour propre*, which had been bruised by a combination of my ill-concealed irritation and his dilatoriness, he kept me waiting an inordinately long time before issuing the per- mits allowing me to capture and export the rabbits.

*   *   *

In dealing with bureaucracy and petty bureaucrats there is only one way to achieve results: be as cool as a glacier and, like a glacier, move forward inch by inexorable inch until you get what you want. But bureaucrat wrestling – like wrestling with alligators – takes time, strength and courage and on occasions you have not the time, for the matter is too urgent. Furthermore, it is not always the bureaucrats who cause problems. We all know from Charles Dickens' Mr Bumble that 'the law is an ass', but who has not encountered the law when it was determined to prove that not only was it an ass but a monumental, moronic, mentally retarded ass as well.

Lack of both time and legal common sense characterize the case of the Dusky seaside sparrow, a case so imbecile, so ludicrous that if it were not for the awful outcome of the whole affair, it would have created roars of incredulous laughter in any after-dinner speech and people would have congratulated you on your powers of exaggeration and sarcasm.

The Dusky seaside sparrow was – and I use the past tense advisedly – a pleasant little bird, blackish in colour with a yellow speckling and a pretty song. It had a limited range in the coastal salt marshes of Florida but gradually, as this habitat was drained and degraded by man, the sparrow population dwindled until there were only five birds, all males, left in existence. These were brought into captivity and a final search for a female was instituted, which proved to be futile. Thus this tiny coterie of male sparrows was the last of their kind in the world. However, nearby lived a close relative, the Scott's seaside sparrow,* and it was suggested that female Scott's could be crossed with Duskies and the offspring back-bred until you had a bird which was so close, genetically speaking, to the Dusky as to be indistinguishable from the real thing. This seemed a sensible approach to a desperate problem, and everyone agreed that it was well worth a try. Every-

* It was discovered, some years later, by analyzing the DNA of seaside sparrows, that the Scott's was not as closely related to the Dusky as some of the other forms of seaside sparrow. This fact, however, does not alter my following comments on the sparrow case, but it does show that the conservationists and the DNA scientists should have got together earlier.

body, that is, except the US Fish and Wildlife Service, the government organization in whose care the remaining Duskies were, and whose job it was to see that the bird did not become extinct.

The argument was not one of finance, for the whole project was to be funded by monies already raised from outside governmental sources (some from our own American Trust), so no federal funds would be involved. No, this argument was a purely legal one, summed up by these extracts from a Florida Audubon Society press release, which turned out to be a cry in the wilderness:

FOR IMMEDIATE RELEASE

GOODBYE FOREVER DUSKY SEASIDE SPARROW?

MAITLAND: The Florida Audubon Society today issued an urgent plea for federal approval of a cross-breeding ploy that would prevent the genes of the Dusky seaside sparrow from being lost to the world forever.

Peter Rhodes Mott, president of the Florida Audubon, said the breeding program should begin now, while there are still five male Duskies alive in captivity. . . . 'We can preserve a breeding population of sparrows whose genes are essentially the same as the Dusky's. We have already raised the money for the first year of the project.' Mott said, however, that the Fish and Wildlife Service has up to this point refused to approve the cross-breeding program. Their lawyers have determined that the offspring of a Dusky–Scott's cross can never be 'pure' Dusky and thus cannot be considered an endangered species. This means the agency cannot spend federal money earmarked for endangered species on the breeding program and that it will not offer protection for Dusky–Scott's crosses released into the wild.

'If the service decides that preserving the Dusky's genes isn't worth the cost, when measured against the need for money to protect other more glamorous endangered species, we can live with that decision,' Mott said. 'But we shouldn't let the lawyers quibble the rare Dusky into oblivion, especially when you consider there is a reasonable and viable alternative.

And we also shouldn't let this species disappear entirely because of the Fish and Wildlife Service's failure to overcome its own bureaucratic inertia. In short, we just can't stand by and do nothing while the remaining Duskies die off, one by one. Especially when an alternative exists and the funding for it from the private sector is available.'

But this battle was not over. Everyone now got into the act: the International Council for Bird Preservation, Dillon Ripley, secretary of the Smithsonian, no less, and no mean aviculturist and ornithologist himself, and Dr Hardy, curator of the Department of Natural Sciences at the Florida State Museum. Dr Hardy wrote on behalf of the scheme to the director of the US Fish and Wildlife Service in Washington. Here is an extract from the reply he received, which shows the craft of gobbledegook raised to a new art form. My translation and comment on this bureaucratic dialect are in italics.

. . . Although your proposal raises some interesting possibilities, we do not believe hybridization of Duskies with similar seaside sparrows is warranted for the following reasons:

1. There is no assurance that hybridization will produce Dusky-like seaside sparrows that will accept the salt-marsh habitat used by Duskies, nor is there assurance that the hybrids will be fertile. (*These crosses have never been tried, so obviously you don't know whether they will work.*)

2. Hybridization will result in a permanent dilution of the Dusky deme, which we consider undesirable. (*'Deme' in the biological jargon means 'a cluster of individuals with a high possibility of matching with each other'* ... *with only four males left, how can you talk about a deme, much less diluting it, much less comment on its desirability?*)

3. There is little information available which indicates that 'back-cross' hybridization is feasible and will produce almost 'pure' Dusky seaside sparrows. (*Again, you haven't tried and what is there to lose, anyway?*)

4. We do not feel that the purposes of the Act can be extended to utilize hybridization as a conservation tool to recover listed species. (*What I think he means is 'we believe in doing nothing rather than something, because doing something may create a precedent and we might all have to work.'*)

5. Approval of this project would set a precedent for hybridization which we cannot support. (*A precedent, once created, turns into a ravening monster, seeking whom it may devour. It is a Pandora's Box, better kept safely locked. Sparrows are expendable, bureaucrats are not.*)

I regret that we differ on the approach to be taken in preserving the Dusky seaside sparrow. Nevertheless, we sincerely appreciate your proposal and hope that you can support the effort we have now planned. (*The 'effort' was to keep looking for a female Dusky!*)

Sincerely, etc. etc.

One commentator described the situation even more caustically than I:

Reasons given for the denial involve the metaphysics of legal quibbling and fear of 'setting a precedent'. Because extinction is a precedent setting event that lasts for ever, we fail to appreciate or understand the legalistic jargon and bureaucratic stumbling of people (some of whom once were biologists and conservationists before they became ensnared in administration) who should know better. It is a bit ironic that the *coup de grace* to the Dusky seaside sparrow population has now been delivered by the very federal agency charged with the job of preventing it from going extinct.

Meanwhile, while the battle raged, the sparrows were getting older and dying one by one. Still the Fish and Wildlife Service remained blandly bureaucratic and budged not.

On 16 June 1987 the last Dusky seaside sparrow died. Thus a bird numbering over 6000 in the 1940s had, in forty odd years,

vanished from our world through the rapaciousness and thought-
lessness of mankind coupled with an idiotic bureaucracy dictated
to by that unpleasant species, the lawyer. An extremely good
young journalist writing on the demise of this bird in the *Orlando
Sentinel* ended on this note: 'Like the canary who warned miners
that oxygen was low, the extinction of the Dusky seaside sparrow
sends a message to us: We are all in peril.'

So these are some of the situations we've encountered on the captive-breeding front in the conservation battle. Sometimes your
enemy may be one small government department, sometimes just
an individual (but this does not necessarily make your task any
easier). On other occasions you find the conservation battle raging
far and wide across the world involving so many diffuse people and
organizations that you begin to despair of ever accomplishing anything. The case which immediately springs to mind involves the
international traffic in wild animals and wild animal products.

In the days when I had just started the zoo in Jersey, I was at a
conference where I was introduced to a Dutch animal dealer, a man
of great charm who spoke exquisite English but was utterly lacking in morals, as I suppose most dealers are. In those days, of
course, there was no real public awareness (as there is today) of the
plight of wildlife, much less international laws protecting wildlife.
Of course, certain countries had what I have always called 'paper
protection' for some of their native species, but the enforcement
of this was, for the most part, desultory and, even when carried out
with some attempt at vigour, was generally abortive owing to the
lack of funds for the necessary manpower and equipment. Also,
the dealers (like today's drug smugglers) were constantly thinking
up new and devious methods of evading the law.

Late at night at the end of the conference the Dutch dealer
sought me out, for he had just discovered that I had started a zoo
and wondered if he could sell me anything. To his disappointment
he had nothing I wanted or could afford if I had wanted it. However, being a convivial fellow, he sat up into the small hours and

became, under the application of alcohol, even more convivial and somewhat indiscreet. Needless to say, having a better head than he had, I hoped to gain knowledge and so plied him with more drink and innocent questions.

'My dear Gerry,' he said, for we had by that time reached the Christian name stage, and he was under the impression that we were blood brothers, 'if you want me to I can get you any animal in the world, whether it is protected or not.'

'I think you're boasting,' I said, smiling at him as if chiding a naughty child, 'that's too much for me to believe.'

'No, no, is true, Gerry, I assure you,' he said earnestly. 'I swear upon my mother's grave.'

'I'm sorry to hear that your mother is dead,' I said.

'She isn't,' he replied, dismissing the quibble, 'but I swear on the grave she will eventually occupy. Go on, test me.'

I thought for a moment.

'Komodo dragons,' I said, knowing these, the largest of all lizards, to be strictly protected in their island homes by the Indonesian government.

'Poof!' he said, gulping at his drink, 'can't you think of something more difficult? Komodo dragons are no problem.'

'Well, how would you do it?' I persisted, genuinely interested.

'You know,' he said, wagging a long, beautifully manicured finger at me, 'the Indonesian government has a launch that patrols the waters around the island of Komodo, eh?'

'Yes,' I said, 'it's an anti-smuggling and anti-poaching patrol.'

He nodded and closed one large moist brown eye in a prodigious wink.

'D'you know how fast she goes? he asked, rhetorically, 'she goes at fifteen knots maximum.'

'So?' I asked.

'So I have a friend on a neighbouring island who has a launch that does thirty-five knots. So we go to Komodo and my friend drops me there. Of course, we bribe the locals for they are terribly criminal people. Three days we catch dragons. My boat comes back and picks me up. Five times we have been chased by the customs launch, but they cannot match our speed. So, poof, dragons for Europe, dragons for America.'

He sighed with immense satisfaction and drained his glass.

'All right,' I said provocatively, 'I'll give you something more difficult. How about if I wanted a Giant panda?'

I felt sure that this would prick the balloon of his conceit. He merely gave me a withering look.

'Why don't you ask me something difficult?' he said, 'not these stupid things. A panda is easy.'

'Well, how would you do it?' I asked.

'Simple. Catch your panda, dye it all black and bring it out legally as a bear. No customs officer would know the difference.'

I went to bed.

At the time of my conversation with the devious Dutch dealer, the international trade in wildlife was carried on with virtually *no* regulations, regardless of the paper or real protection for animals in their own countries, which in either case was inadequate and hard to enforce. The attitude to the fate of the various species involved was callous beyond belief. Tigers, spotted cats, crocodiles and sea turtles were threatened because the fashion trade relied on their skins. The numbers of monkeys and apes from the tropics dwindled because medical institutions in America and Europe wanted them for experiments, hundreds of thousands of birds, reptiles, amphibians and fish were being taken from the wild for the pet trade, few surviving the well-meaning but ignorant ministrations of their new masters.

However, since the early days things have got a little better. After ten years of research into the animal trade the IUCN proposed the Convention on International Trade in Endangered Species (CITES). This was signed in 1973 by twenty-one nations, and to date over ninety countries have signed. The purpose of this Convention is to be able to monitor and regulate the international trade in wild plants and animals and their products – for example, ivory, furs or skins – and to protect those species threatened by it.

The point of the Convention is not to deny countries the foreign exchange which wise exploitation of their wildlife brings, but to make sure it is controlled efficiently and therefore sustainably. Thus CITES is better than nothing, but there are still loopholes through which unscrupulous traders and dealers can slip, and one of these is the fact that, though one country may have signed the Convention, it is possible that its neighbour (or neighbours) has not and this opens a conduit for smuggling, for an animal or plant coming out of a country *not* signatory to CITES cannot be impounded by the customs.

A complex situation arose from this very fact over the illegal trade in the Golden-headed lion tamarin from Brazil. This minute and beautiful primate had been whittled down in numbers by capture for the pet trade and the fact that their forest homeland was being felled at a ferocious speed to make room for farmland and cattle grazing. Suddenly, to everyone's astonishment, a Belgian animal dealer offered twenty-four for sale. These, it was thought, could have represented up to a quarter of the world population and to have them appear in the pet trade was appalling. Further investigations revealed more animals in zoos or private hands in Japan, Hong Kong, France and Portugal. Altogether, fifty-four tamarins were located. It became apparent that all these animals had been illegally captured in Brazil where they are fully protected, for Brazil is a signatory of CITES. Once caught, the animals were smuggled into Guyana and there sold by animal dealers. The fact that the animal did not come from Guyana and the likelihood of finding wild ones there was as probable as finding a colony of Polar bears in the Sahara Desert made no difference. At the time, Guyana was not a signatory of CITES, nor was Belgium for that matter, and these countries could do what they liked. Thus possibly as much as half of the world population of this delicate little animal was in private collections or at zoos.

Immediately the forces of conservation set to work. To begin with it was unthinkable that such a high proportion of the world population of such a creature should be scattered about the globe in this fashion. Also, the animals had been illegally caught and smuggled out of their native Brazil. It was imperative that they were returned to Brazil if possible; if this could not be done, then the creatures should be kept together so that they founded the nucleus of a captive-breeding group. This was easier said than done. Jeremy was in the thick of all this since he has a deep and abiding love of the marmosets and tamarins. The first step was to get the animal dealer in Belgium to relinquish the animals in his possession. Not unnaturally, he was reluctant to do this since the animals represented considerable financial outlay for him and if he returned them to the Brazilian government he would get no profit. By this time the IUCN, the World Wildlife Fund, the Brazilian government, the Belgian government, the National Zoo in Washington and ourselves were all involved. We were all quite determined not to buy the animals (which we could have done) as this would have looked as if we were condoning and, indeed, encouraging the illegal trade in these creatures.

Representations were made at top level. The Belgian authorities were asked by the Brazilian government to use their good offices to get the animals repatriated and the Duke of Edinburgh, as President of the WWF, also wrote to the Belgian authorities. At length, to our relief, this public pressure had its effect and the animal dealer reluctantly agreed to send all but eight of the animals back to Brazil. Following this good example, the specimens in Japan were also sent back to their homeland and the titles to at least some of the animals held elsewhere were given back to the Brazilians. Then another problem arose, for the Brazilian authorities – not being versed in the complexities of conservation – could not understand why we did not simply release the animals which had been repatriated back into the nearest piece of forest. Of course, to release animals used to captivity back into a strange forest area would be tantamount to killing them. We at last persuaded the Brazilians that the animals should go to the Primate Centre at Rio de Janeiro to become the foundation of a captive-breeding group. This was duly done and in due course two further groups were founded, one in Washington and one in Jersey. While all these colonies are flourishing, we are making strenuous attempts with the aid of the Brazilian government to save some remnants of the forest so that, when the numbers in captivity are sufficient, a reintroduction scheme can be worked out for the species as has been done for its relative, the Golden lion tamarin, a story I take up in Chapter Five.

CITES is an enormous step in the right direction – one it would have been unthinkable to propose twenty-five years ago – and yet the volume of wildlife trade, both legal and illegal, has reached astronomical levels. In the five years after CITES was established the so-called legitimate imports of wildlife products to the USA alone rose from four million 'items' to 187 million 'items'. Just three years later that trade was worth a billion dollars! Over twenty million butterflies are exported from Taiwan each year to end up as dried and desiccated specimens 'enhancing' the beauty of suburban walls all over the world, and hundreds of thousands of sea creatures are killed each year so that their shells may gather dust on suburban mantelpieces.

The world's inexhaustible appetite for ivory has devastated the African elephant and, because the horns of the rhinoceros are worth their weight in gold, the numbers of this wonderful antediluvian creature have dwindled to a pathetic few thousand. It is big money which urges on the rapacious exploiters, as in the drug trade. If the coat of an ocelot fetches $US40,000, what use are the creatures except dead? Nine birds of prey recently smuggled into Saudi Arabia were sold for $US200,000. Faced with this sort of money being legitimately or illegitimately earned from wildlife, the sums put into its conservation seem pitifully small, and the heart of the conservationist falters. Given the volume of trade and the economic pressures felt by some countries, it is no wonder that the CITES loopholes are gaping wide open.

Even if a country signs the Convention, it is in no way legally bound to obey the legislation which may be proposed by other countries: it can declare exemptions to the legislation in its own interest. Another point is that harassed customs officers are not supposed to be zoologists and so have difficulty in spotting the illegal species on the Convention's list. Furthermore, in some cases an extremely rare creature can look very like a common related one, so that only an expert could tell them apart. Another major problem is what to do with the animals if they are discovered by customs and confiscated? They cannot simply be sent back to the country of origin and released, and generally in the country of origin there are few, if any, people with the expertise for looking after these creatures. So what do the customs do? Almost inevitably they have to send them to a zoo or similar institution. We were asked to help in a problem of this kind a few years ago.

Madagascar has several different species of tortoise, all rare and all given at least paper protection. Of these one of the largest and most handsome is the Radiated tortoise. A full-grown specimen measures about two feet long and weighs about forty pounds. On the honey-coloured carapace is a vivid black star pattern. We already had a number of these handsome reptiles in Jersey and they were breeding at several other places in Europe and America, so the species seemed to be safe in captivity, but not in the wild. In spite of being protected, it is still eaten by some Malagasy people and of course the terrible fires (man-created) which decimate Madagascar's remaining forests take their toll of slow-moving creatures like tortoises.

We were somewhat ill-prepared and taken aback one day to receive a telephone call from a senior officer in the UK Wildlife Division of the Department of the Environment. What, he enquired, did we know about Radiated tortoises? We said they came from Madagascar and that we kept and bred them. He said this was excellent news as he hoped we could help out in a little problem he had. He had been contacted by the customs authorities in Hong Kong who had spotted some Radiated tortoises being smuggled into the colony and had quite rightly confiscated them. But now they had the problem of nowhere to put them, and could we help out? Slightly dazed, we said we would certainly try to help and how many tortoises were there? Sixty-five, said the man with relish. They had been smuggled out of Madagascar and were destined for Chinese kitchens, to be turned into savoury stews, casseroles, pies and similar delicacies. Having said we would help, of course we had to, so we cleared out a whole room in the Reptile House. Sixty-five Radiated tortoises duly arrived, some only the size of a saucer, some like footstools. Most of them were in good condition but some were suffering from malnutrition and general ill treatment. Four died within a few days of arrival, but the rest flourished. It was an arresting sight to open the door of the room and see the floor literally cobbled with the shells of these lovely reptiles.

It was, of course, impossible to return them to Madagascar, but we informed the Malagasy authorities, with whom we were collaborating on other conservation matters, and they asked us to solve the problem. As soon as the animals were fit and healthy we sent males and females, on a breeding loan arrangement with the Malagasy government, to other zoos that already possessed breeding groups, where they proved to be welcome new blood lines. Looking back on this incident, I suppose we should consider ourselves lucky; instead of sixty-five tortoises it could have been fifty Komodo dragons or a hundred or so leopards, or even a clutch of elephants.

The key to the prevention of this terrible rape of wild things, both plant and animal, is education. People must be taught that all things in nature are an endless, renewable resource if used wisely and not squandered. If they are taught that their natural heritage is something to be proud of, to be guarded and not wasted in selfish, short-term gain, wise utilization of nature follows to the benefit of all. A case in point is our involvement with a beautiful bird from the Caribbean, the St Lucia parrot, a gorgeous creature clad in green, red, yellow and blue feathering. When its plight came to the attention of David Jeggo, our Curator of Birds, some fifteen years

①Les Augres Manor, some of which was built in 1530. Our flat occupies the top two
floors of the central part.

aptions to photographs:

②Our two male Przewalski horses. Jersey Zoo acts as a 'stallion station' for the world's
captive breeding herd, now numbering more than 700. Extinct in the wild, the
Przewalski horse will be reintroduced soon to its native Mongolia.

(3) The picture of the baby gorilla that won over Princess Grace.

(4) The ebullient Simon Hicks, Trust Secretary and fund raiser par excellence, whose vocabulary does not include the word 'impossible'.

(5) Tony Allchurch, for many years our veterinarian and now also our General Administrator.

(6) Chumley and Lulu, the two chimpanzees who took tea with my mother.

(7) David Niven, who was best man at the wedding of N'Pongo and Nandi to Jambo, with an appropriate bouquet for the brides.

8) A Ring-tailed lemur, one of the lemurs that led me to Lee.

9) Lee.

10) One of the Radiated tortoises we rescued, apparently singing 'Alleluia'.

11 Jeremy Mallinson, who came to the Jersey Zoo 30 years ago looking for a temporary job and is now its much loved and respected Zoological Director.

12 Les Noyers, the farm that fell into our lap.

13 Our international training centre brings people from all over the world.

14 Sending home a pair of St. Lucia parrots, bred in Jersey, accompanied by the Prime Minister of St. Lucia. This is the first time I have had an opportunity of holding something over a prime minister.

15 Our Patron, the Princess Royal, being introduced to one of our trainees, Kanchai Sanwong of Thailand, by David Waugh, our training officer.

16 The Pink pigeon of Mauritius. Fewer than ten wild birds are left, but more than 200 have been bred in captivity. We have used some of these in a reintroduction programme which has, so far, been very successful.

17 John Hartley, wrangling a Round Island boa with Lee and me. As well as being my Personal Assistant, John masterminds all of our diverse work in the Mascarenes.

18 The Golden lion tamarin of Brazil, saved by a truly international co-operative effort, involving governments, scientific institutions and zoos.

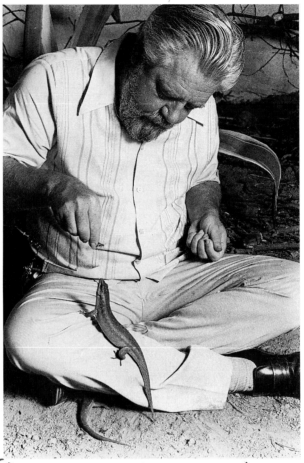

(19) My first visit to Round Island. The Telfair's skinks were so tame, like many island animals, they would come and sit in your lap and share your picnic.

(20) My gift from the staff on the twenty-fifth anniversary of Jersey Zoo, a silver matchbox filled with gilt scorpions, to remind me that my boyhood 'matchbox menagerie' had turned into something rather special. End of Captions

ago, this parrot, once common, had sunk in number so there were only just over a hundred in the wild state and a few languishing in captivity in small cages, kept singly as 'pets' and so unable to breed. This had come about by a conglomeration of different factors: the disappearance of habitat by destruction of the forest, shooting, since Jaquot (as the bird is known in St Lucia) is considered a welcome dish at any time of the year but especially on Christmas Day, and a steady smuggling trade in young birds to parrot fanciers in Europe and the USA.

With the blessing of the St Lucian government we were allowed to collect seven young birds (the only St Lucia parrots legally allowed outside the island), and bring them to Jersey in the hope of founding a breeding colony. As with all our rare creatures, the birds remained the property of St Lucia. They settled down well and we had high hopes, but we did not expect immediate success as parrots, for the most part, are slow breeders. In the meantime, Gabriel Charles, Chief of Forestry, and the Ministry of Agriculture were making valiant attempts to save the remnants of forest on the mountains of the island, for they not only form a vital watershed for the people of St Lucia but are the parrots' last stronghold. A total ban on hunting all over the island and a census of all parrots held in captivity were other most helpful steps in the right direction. They also employed a young Englishman called Paul Butler to help them with the campaign to save the parrot and he was well aware that education was the key to success.

From us he got huge colour posters of the parrot which he put up in schools, government departments, even in shops and bars. He had got the far-sighted government to make the St Lucia parrot the National Bird, and he published booklets for schools about the adventures of Jaquot and the importance of its forest home. Within three years there was not a soul on St Lucia who did not know about the parrot, that it was their National Bird and should therefore be protected. Protecting it, of course, meant protecting its forest home and the watershed of the island.

Then disaster struck in the form of Hurricane Allen, which

laid to waste large areas of forest and we feared the worst for the wild parrots. The huge trees felled lay criss-crossed like a spilt box of matches and it was impossible to get through the tangle to count the dead and help the living. A frantic message for help came to us, and within twenty-four hours we had David Jeggo on a plane armed with a huge chainsaw. Fortunately, David found that though the destruction of the forest was bad, the loss of birds was not as great as we had feared. In fact – and this is a testament to the work of Gabriel Charles and Paul Butler, now known fondly all over the island as 'Paul Parrot' – the parrots which had survived the hurricane but were weak and starving were being picked up, cared for and handed over to forestry officers by the St Lucians themselves! Gabriel told me later that if the hurricane had happened *before* the education programme the people would have collected the helpless parrots to eat them.

Meanwhile our breeding programme in Jersey was going full swing and we had bred fourteen of these lovely birds. It was time to start thinking of sending some back so that the St Lucians could start their own captive-breeding colony. The Trust gave the Forestry Department a grant to build the necessary aviaries and David Jeggo was sent out to help with construction and design. Two of our parrots at the right breeding age were chosen, but we had to make sure of their sexes. There are many species of bird like the St Lucia parrot in which the sexes look identical, and before the invention of a wonderful piece of equipment you could have two males or two females languishing together while everyone wondered why the poor things did not breed.

The equipment is called a laparoscope and it was, of course, invented for human medicine but has proved invaluable for veterinary work. It consists of a powerful light source, a flexible light cable of glass-fibre elements, which is about the thickness of a pencil, and a precision-made viewing piece called an endoscope. The whole set-up allows you to examine the interior of a human or any other species of animal under light anaesthesia, and with the minimum of trauma to the patient. In the case of birds a minute opening is made under the wing, an incision just large enough so that you can insert the laparoscope. Then you can move it around gently, moving aside the various organs until you have a clear view of the sexual parts, two oval testes in the case of the male and a cluster of ova in the case of the female which looks remarkably like a bunch of grapes.

Our next step, when the aviaries in St Lucia had been built, was to suggest that the Prime Minister himself should come over to take possession of the birds. Simon Hicks, our Trust Secretary, was despatched to St Lucia with the invitation, and the Prime Minister said he would be delighted if we could fix a suitable date. We then got on to British Airways, who have an excellent scheme called 'Assisting Nature Conservation' by which (when they can) they fly, for free, animals, equipment (and sometimes bodies) to various parts of the world where we have conservation programmes, such as in Mauritius or Madagascar, and they do this for other conservation organisations as well. So we contacted them and asked, tentatively, if they could help us ship something from the Caribbean. Thinking it was some animal they said they would see what they could do. When we divulged that it was the Prime Minister of St Lucia, they were somewhat taken aback but rallied around when we explained the importance of the situation, and agreed to waft the Prime Minister and his entourage to Jersey.

The great day dawned. Unfortunately, I had been ill and so – on doctor's orders – I could attend only the handing-over ceremony. We had a dais built outside our breeding aviaries, with microphones so that the Governor of Jersey, Sir William Pillar, the Prime Minister and I could all make speeches. The weather, unfortunately, was grey and overcast with a slight but persistent drizzle, but this did not deter the crowds or the press reporters and TV crews. At last the entourage came sailing majestically down the main drive. Two enormous cars, glossy as whales, were flanked by police outriders immaculately uniformed, blue lights flashing. It was most impressive. As the Prime Minister alighted with Sir William, the children from Trinity School in our parish sang the St

Lucian national anthem, which I think surprised and touched the Prime Minister, who stood to attention until it was over. The children had been practising for weeks and their voices sang the lovely anthem very sweetly. The parrots in the aviaries behind us were thoroughly over excited by all this attention and squawked and screamed and almost drowned out the speeches. As the Prime Minister was making his acceptance speech, it began to rain quite heavily, so I had to hold a huge red-and-white golfing umbrella over his head. It was the first time I had ever held an umbrella over a Prime Minister's head, but I find a great charm of life in conservation is the novelty of the things you have to do.

The ceremony over, the Prime Minister and his wife were taken on a short tour of the Trust's headquarters and then came back to the manor house for tea. After they had left, Paul Butler stayed behind to discuss conservation matters and he told me this story, which I think exemplifies the great impact that education can have on saving a species.

An American gentleman arrived at the airport in St Lucia, got into a taxi and asked to be taken up to the parrot forests. The taxi driver became suspicious for some reason best known to himself and was convinced that the man was up to no good. Dutifully, however, he drove the unsuspecting American up into the forest to a place where he might conceivably see parrots and left him, promising to return some hours later. Then he drove as fast as he could to the nearest telephone and got in touch with the Forestry Department. He vouchsafed to them the details of the parrot smuggler, for he had heard tales of these men who dope the birds and then pack them in layers in the false bottoms of their suitcases. He had not a jot or tittle of proof, but the Forestry Department took his report very seriously. It presented them with a nice problem in diplomacy. The man was an American and, like all the islands in the Caribbean, St Lucia relies heavily on the USA for its tourist trade. To stop an American citizen and search his luggage on suspicion of smuggling parrots would hit the headlines and, if the man were innocent, cause a tremendous furore. After some thought, the Forestry Department came up with an extremely wily plan. They phoned the FBI in Miami and explained their dilemma. Could the FBI help? The FBI sure could, and would, and came up with a wily plan of their own. They got the man's name and return flight number from St Lucia, and when the flight landed in Miami the FBI announced that there had been a bomb scare and that all passengers' luggage would have to be searched. Of course, the only luggage that was searched belonged to the alleged parrot smuggler. As it happened, they could not find so much as a parrot feather in his belongings. But it was nice to know that this whole chain of events had been put into operation by a taxi driver who – in the days before the Ministry of Agriculture, the Forestry Department and Paul Butler had started their campaign – might well have said to the American 'What parrots?' or, even worse, 'Do you want me to help you catch some?'

4

The Jigsaw Strategy

Over the years, Jeremy has always referred to the Trust's 'multifaceted approach' to captive breeding for conservation and, although we tend to pull his leg over this oft repeated and somewhat ponderous statement, it is perfectly true that this phrase sums up our work and the reason for our success. The Trust's multifaceted approach is really made up of three stages, each different but each, as it were, interlocking with the other two bits of a jigsaw.

Stage one is the choosing of the species which you think will benefit most from your help, setting up breeding colonies in Jersey and then, when there are sufficient numbers, forming 'satellite' colonies in other zoological collections of good repute, usually in Europe or America. At this point you can say with some degree of truth that, leaving aside disasters, you have saved the species in captivity. Stage two is the establishment of colonies in the species' country of origin, for that after all is where endangered species *should* be bred. The climate is right and the appropriate foods are available and, most importantly, the people can see their own animal treasures and learn to appreciate them. The third stage is the arduous task of releasing into the wild animals bred in Jersey or our satellite colonies or indeed in their home country. This final stage is a story in itself and is the subject of Chapter Five.

It became apparent early on that there was no point in even stage one of our activities unless we had the involvement of the government of the country where the animals come from. Over the years, I have become distressingly aware of how little is known about conservation and the necessity for it by those people in the position of power to make decisions: the politicians. I have a little black book in which I write down the more inane statements made by world leaders so that, when I complain about governments and people think I am exaggerating, I can quote them chapter and verse.

Top of my list, of course, is ex-President Reagan. His grasp of ecological problems and his grave concern for the environment are, I think, nicely encapsulated in these two statements: first, that trees cause pollution and, second, that it did not matter if the redwood forests were hacked to pieces for, as the president sagely remarked, when you've seen one redwood you have seen them all. My next prize for the oafish or mentally retarded statement must surely go to the Indian minister who, when faced with conservation opposition to the building of a dam which would flood a very important piece of forest and its wildlife, said angrily that 'we cannot afford these ecological luxuries'. When ecology becomes a luxury then we are all dead.

The Minister for Mines in Queensland, Australia won my heart with his remarks to the Press when there was controversy over the proposal to drill for oil on the Great Barrier Reef. First of all, he said, there was no such thing as an oil slick. He then added that if by some billion-to-one chance oil should escape, there was no problem for (and here I quote him verbatim) 'as every schoolboy knows, oil floats on the surface of the water and coral lives underneath the surface, so no harm could be done'. One wonders at the standards in the school the minister went to, always providing that he went to school. A high-ranking Brazilian official was reported as having said that there was no 'proof' that chopping down forests could alter the climate. When you consider the vast areas of desertification brought about by the blind policy of forest eradication, one wonders what sort of biology lesson the minister received in *his* school.

When I was in New Zealand, I had – for my sins – to lunch with the entire New Zealand Cabinet, a harrowing experience, made all the more so by a minister trumpeting on about the Wildlife Service wanting to eradicate some feral sheep on an island in the south, only the second known nesting colony of a rare species of albatross. He said the idea was ridiculous. He had been a sheep farmer all his life and he had never known a sheep tread on an egg or a bird nesting on the ground. I said I thought that probably the necessity for getting rid of the sheep was because they were increasing in numbers and, by over-grazing, probably making the habitat unsuitable for the albatross.

'That may be the reason,' he conceded, 'but what's it matter if the albatross leave this island? They're so far south, no one can get there to see the bloody birds.' I said there were a great number of Rembrandts in the world which I should never see, but I would not suggest burning them on that score. He made no reply. I think that perhaps he did not understand me. He probably thought a Rembrandt was a make of beer.

With this sort of experience behind me when it came to the time for us to work with governments in our endeavours, I viewed the prospect with a certain alarm and despondency, but our dealings have been for the most part very straightforward.

As I was writing this, however, I had a discussion with Jeremy about a project we were going to fund in a country which shall be nameless. I asked him why were we being so laggardly, why had we not gone ahead? 'Elections,' said Jeremy dolefully. 'The party in power, as you know, is in favour of our proposals. However, they are coming up for elections in a month or so, and if the other party gets in they will probably throw out any decisions made by *this* party, so we thought it prudent to wait.' So a vital piece of conservation work could not proceed until little politicians had done all their strutting and fretting and their wildlife had slipped one notch nearer to oblivion.

But on the whole we work very well with governments. I think the chief reason for this is the Accords we have set up for both parties to agree and sign, and they have proved invaluable. They are signals of our good intentions that all the creatures and their progeny remain the property of the country of origin and are recallable at any time. This proves, if proof were needed, that we are not simply helping ourselves to rare animals but are working with the government for the benefit of their fauna. Thus the governments concerned need have no fear that we are 'colonialist robbers of endangered species'. A second reason these Accords are so valuable is that they set out on paper the ways and means of implementing stages two and three of our multifaceted approach to saving species.

As our breeding successes in Jersey became greater, it was stage two, the establishment of *in situ* breeding colonies, that gave us the most cause for concern. Although it was obvious that where these animals *should* be bred was in their home countries, there was generally no one there with the expertise to undertake such a task, even supposing we gave a grant to set up a captive-breeding centre. It was clear that we should have to train people for the task and this meant bringing them to Jersey. So we planned to build a training centre, a complex we would come to call our 'mini-university', which contained student living quarters, a lecture hall, a library and many other necessary facilities. It was going to be a huge and costly building and, although I was loath to cover some of our lovely farmland with cement, there seemed no other way for the Trust to fulfil its function. We got the architect's plans, pored over them, argued over them and then got a revised version on which we hoped we had corrected our mistakes and not overlooked anything vital.

Once again, I picked up my begging bowl and went to the United States, and with typical American generosity the money was readily forthcoming. From another source in the UK came the all-important grant to allow the creation of scholarships for people from what are laughingly called 'emerging nations', who would not have the wherewithal to travel to Jersey for training. So we had the finance for both the building and the scholarships. I was not, however, happy with the building. Owing to limited funds, we could not afford frills and furbelows and, in spite of our excellent architect's efforts to soothe me, I still felt it was going to look somewhat like a cement shoebox. Having seen what horrors have been perpetrated in zoos all over the world in the name of the Goddess Cement, I tend to view this useful substance with a certain distaste. However, there was nothing to be done about it – or so I thought. Then something very curious occurred, which showed that Fate was on our side.

For many years a certain Mrs Boizard had come to our flat twice a week to sweep through it like a tornado and leave it sparkling bright. Mrs Boizard's younger daughter, Betty, had come straight from school to work for the Trust and, over the years, had taken over the zoo accounts office and now rules it with a rod of iron. When Mrs Boizard came to the flat, Betty would always pop up and have five minutes' gossip with her mother. On this particular occasion, Mrs Boizard said, 'I see Leonard du Feu has put his house on the market.' Betty looked incredulous, as well she might. Leonard was our nearest and most long-suffering neighbour, never complaining when animals made strange noises at night, not even protesting mildly when our South American tapir, Claudius, escaped, trampled a field of just blooming anemones into a pulp and then proceeded to Leonard's garden and broke all his cloches. Leonard was a neighbour we prized above rubies. His Jersey property had been in his family for ever (something like five hundred years) and his fields, as they say, marched with ours. His house lay two minutes' walk from the manor and was really three houses in one, with a small worker's cottage, a huge coolroom and massive granite outbuildings. We had never, in our wildest dreams, imagined that Leonard would sell the family home, but once children are grown up and have moved away a house that is three houses in one takes a lot of upkeep.

Breathlessly, Betty descended to the office and gave the news to the astonished John Hartley, who immediately phoned me at my house in Provence, where I was busy writing a book. I told John to get on to Leonard and just hoped and prayed we were not too late. When John got hold of our erstwhile neighbour he said that the property had been on the market for a month and he could not, for the life of him, imagine why he had not thought of us as possible purchasers. On the face of it the whole thing sounded straightforward, but it proved to be anything but, for since the days of setting up the zoo times had changed. It took no less than three States of Jersey Committees to give their blessing before we could purchase the property and put it to the use we wanted. The States had just passed stringent laws relating to the ways in which properties (particularly agricultural properties) were used and, needless to say, none of the purposes for which we wanted to use Les Noyers fitted any of the categories delineated by the law. However, Jersey is proud of us and the work we do, and on this occasion (as they had done on previous occasions) they showed a massive vote of confidence in us. I do not say for one moment that the law was bent, but finally it was perhaps a little out of true and Les Noyers was ours.

To say we were delighted would be wholly inadequate. Instead of the massive concrete block we were contemplating for our training centre we had an elegantly beautiful old Jersey farmhouse with massive outbuildings and eight acres of land. There were some restrictions as to what use we could put the land to, but this did not matter for it was the house and outbuildings that we really wanted. So, as soon as the property was ours, we started renovations. Part of the house was used as student dormitories and living quarters, part converted into a flat for a housekeeper. We set up an elegant library called the William Collins Memorial Library, for before he died Sir William (my publisher) most generously gave us every zoological and natural history book that Collins had published or would publish in the future. In the huge outbuildings we had a lecture theatre constructed to seat sixty-four, with all the latest audio-visual aids. Above it were offices, a small museum, a graphics and photography area, a darkroom and a video suite, and we had still only taken up one half of the giant granite barn.

Once we had the training centre, of course, we needed a training officer. We knew this would have to be a very special sort of animal, someone biologically qualified but who could deal with a great variety of people from all over the world with tact and sympathy – someone in fact who could be both a father confessor and a father figure. To get the best, we felt we had to advertise the job. Naturally, we were inundated with candidates and from this pile we had to sort out the impossible ones, such as the lady from Penge who loved animals, had fourteen cats and had spent a holiday in Majorca, and the eighteen-year-old schoolboy from Somerset who said he had always liked foreigners in spite of them being different and had always wanted to teach them. So out of the flood we extracted some fourteen possibles and the interviews took place in London.

I know it is probably a nerveracking task to go for an interview, but people who apply for jobs should spare a kind thought for the interviewers, because choosing people sight unseen is a gruelling task. You have their *curriculum vitae* in front of you, but for a job like this you are looking for a special personality, because we are a very small organization and from the point of view of personality we cannot afford a rotten apple in our barrel. Sometimes, if the candidate is nervous, it is difficult to make an assessment. Fortunately, there is a strictly adhered-to procedure as to the length of time that can be spent, the rules being that everyone who sits on the board must not waffle and must be able to drink unlimited quantities of bad coffee without flinching or having it impair their judgment. On this occasion, we were fairly speedy in eliminating candidates, especially the young man who came shambling into the room with his fly undone, waved us all a genial 'good-day' and asked if anyone had a light for his cigarette. We were of the strong opinion that his CV might have been forged. Suddenly, so swift had we been in the elimination process, we realized that we had only one candidate left. There were one or two faintly possible among the ones we had seen, but nothing that had stirred our enthusiasm. John went out to the anteroom and came back with the news that the last candidate had not turned up.

'Well,' I said, 'there's nothing for it. We'll have to advertise again.'

'And this time let's put in a bit about having fifteen cats not being the same as having a PhD in biology,' said John.

'Yes, and that though foreigners are not English, a good training officer must try to overlook it,' said Lee. 'After all, I'm an American.'

'Well, I suggest we all go somewhere to get the foul taste of this coffee out of our mouths with a good, old-fashioned non-foreign drink, like brandy and ginger ale,' I suggested.

Fate has always played a sly game in the history of this Trust. She has waited until the very last minute before coming to our rescue. As we were shuffling our papers together, there was a knock at the door and at our unanimously shouted 'Come in', Dr David Waugh entered, the missing candidate, who had missed his bus, his train and everything else he could miss, but had come nevertheless. We could do no less than give him a cup of cold coffee and proceed to question him as deeply as if we were the CID and he was suspected of being Jack the Ripper. As he answered us, it became apparent that this, the last candidate, was the one we wanted. With light hearts we sent him on his way, congratulated ourselves on our brilliance and found our way to a convenient hostelry.

So David took up his post and his first job was to plan our training programme. This he did speedily and with great flair, because it is necessary for a programme as complex as this to be supple. I had a friend once who used to say of some programme or other, 'What is needed is rigid flexibility.' Although the terms are opposed to each other, they make a wild sort of Alice in Wonderland sense, and so we have always striven for 'rigid flexibility' and this is what David produced. But while he was doing this, we knew that we would have to have (to aid David with his minor United Nations) someone that Americans, with their occasionally deft choice of terms, call a House Mother. Her terms of reference were as stringent as the ones David had to live up to:

she had to control a mixed bag of people from all over the world, love them, be firm with them and, above all, be capable of coping with what you could describe as idiosyncrasies but which were merely differences of understanding because of linguistic failure, different culture or religion or simply the fact that they were far from home and lonely.

Olwyn, if she will excuse me saying so, is my perfect idea of a House Mother. A substantial woman, always impeccable, she exudes the air of a farmer's wife, one who has to cook the meals for ten children, a husband and eight field hands. One who milks the cow, gathers eggs from the hens, feeds the pigs, and is up at dawn or before to bake her own bread. A person, in fact, of great sympathy who could take anything in her stride. And, indeed, she and David have had to.

Olwyn's cooking abilities are fabulous but she has that rare quality of a great cook, to be able to orchestrate her meals to her audience. She has, without batting an eyelid, been able to cope with a student from Pakistan whose great passion was sandwiches of tuna fish, fried potatoes, tomato ketchup and marmalade, and with students who requested sardines and lemon curd or marmite and lemon curd. In the face of such demands, a lesser woman would have quailed, but not Olwyn. She even resigned herself to the Californian who received every week an enormous care package consisting entirely of chocolate and candy of brands that backward Jersey did not have, and the Uruguayan girls who wandered around asking constantly for refills of hot water for their maté pots – which they clung to with the tenacity of babies to their dummies – and one of them insisting on eating no less than fourteen oranges a day.

Nor were Olwyn's problems purely culinary, but only particularly so. We had a student from Nigeria who persisted – quite rightly in my view – in calling Olwyn 'Mama'. He came to her some two days after arrival and said 'Mama, I get terrible burning pain for belly.' Fearing the worst, Olwyn questioned him closely and discerned that he was only acutely constipated. Firmly but kindly,

110

she gave him a suppository and explained how to use it, and that in half an hour he would get relief, so he ought to be within handy range of the lavatory. It shows how, even if you think a person understands you, it is wise to make sure. He came back within the hour and told her that the suppository had not worked.

'What did you do?' asked Olwyn.

'Mama, I go for latrine, I push this up and I pull it down. I do it for thirty minutes as Mama said. Still I have pains in my belly.'

David also had his problems. One day, giving an English lesson to a Malagasy student to improve his rather tenuous grasp on the language, David made up a story in which a man acquired a broken arm. The students were then asked what the man had broken. The Malagasy student, obviously thinking a broken arm too effete, gave as the list of things the man had broken as eyes, liver, lungs, stomach, ear, heart and tongue. When asked why he had listed all these unbreakable parts of the human anatomy, he replied that it was because he remembered the words and was extremely proud of this.

On another occasion, David was asking each of the students their National Bird. Things went well until he reached the student from Ghana who pondered with furrowed brow.

'An eagle, sah,' he said at last.

'What sort of eagle?' David asked.

The student again thought long and deeply.

'An extinct eagle, sah,' he said at last, triumphantly.

Then there were the two students from Thailand who arrived during a particularly savage winter when the snow lay deep and crisp and even. They were fascinated by this commodity, which they had never seen before. It was perhaps unfortunate that most of their studies had to be done out of doors where they felt the cold most grievously. After a week or so they complained to David of having itchy feet. Investigation showed that the moment they got back in the house the first thing they did was whip their shoes and socks off and put their feet against the nearest radiator. It was not to be wondered at that they were suffering from chilblains, but it is also not to be wondered at that chilblains are unknown in Thailand.

I had never thought to see this, our mini-university, come into being during my lifetime, but it seemed no time at all before I was playing croquet on the back lawn of Les Noyers with students from Brazil, Mexico, Liberia, India and China, and that they kindly allowed me to win was in no way the only reason I felt proud. As a matter of fact, we have more applications worldwide from students wanting to come for training than we have room for (which is about thirty a year), and it is heartbreaking for them and for us to have to put them on our ever-lengthening waiting list. We have, quite literally, applications from all over the world, for as well as South America, Africa, Asia, Indonesia and Japan we have students in Europe, the USA, Canada and Australia wanting to come and benefit from our training scheme. These, for the most part, are youngsters who are looking for career training and who can pay their way over and afford our modest tuition fees, so we have to be very careful that they do not swamp the students from poorer countries, for our space is limited. But this mix of people from all nations has worked out very well indeed, I think because our trainees passionately share a common purpose, which is the saving of endangered species. It is good for them to meet one another and suddenly realize, with surprise, that all countries have conservation problems, not just their own, and they can help one another by exchanging ideas and information.

While the broad facets of the Trust's approach to species conservation were being cut and polished over the years, there were many other facets that were being refined. The day-to-day routine in the manor grounds had to be attended to and improved, to keep the whole machinery of our ever-expanding organization functioning. For example, take the veterinary side of maintaining a large collection of priceless endangered species.

Having gone on record as saying that the two most dangerous

animals to let loose unsuperintended in a zoo are an architect and a veterinary surgeon, I am quite inured to architects approaching me with ill-concealed expressions of loathing as if I were Boris the Impaler of Bulgaria, and veterinarians edging round me with all the caution they would afford a mad bull or a rabid dog. I suppose the only comforting thing about this was that it proved they were literate or at least had taken the trouble to have the book read to them.

I remember the first time I asked both our veterinary surgeon and our doctor to attend a sick ape, I caused a considerable amount of embarrassment which I, in my innocence, could not understand. In Europe, among the major zoos, and in America as well, they unhesitatingly made use of the skills of a doctor should they be necessary or, indeed, a dental surgeon. In the work of trying to diagnose and mend the myriad wonderful machines, from man to mouse, there have been three separate meandering streams of knowledge: the exploration of the human being, the garnered knowledge about the function and illnesses of the domestic animal, and the last stream (until recently a veritable trickle), the study of wild animals in captivity. Naturally, the human being was paramount and therefore research in that area was the most comprehensive. But because the three streams were not intermingled, the veterinary sciences lost out. Although learning to apply the increasingly developed skills of human medicine, the veterinary surgeon was taught only about domestic animals and thus viewed the arrival in his surgery of a baby chimpanzee or a giant otter from Guyana with a certain amount of trepidation. Because of this limited approach to the teaching of veterinary surgery, zoological collections suffered.

I remember in my youth attending innumerable post mortems when the animal's corpse was hacked to pieces and the inevitable diagnosis was TB. It never seemed to occur to anyone why so many diverse creatures, from ostriches to antelope, should suffer only from this malady, nor did it seem to occur to anyone that there might be a cure or a preventative. Fortunately, at long last, those days are over. Now there is a free interchange of knowledge and manipulative skills.

It was a few years ago that Tony Allchurch made his appearance in our midst. As an enthusiastic young veterinary surgeon he had taken up a partnership with Nick Blampied who, like his father before him, had given us stalwart service over the years. I think that Tony – like most veterinarians of worth – was fascinated by the wide variety of clients we provided him with, and all the irritating and convoluted methods they employed to make his life intolerable but nevertheless interesting.

One of Tony's first jobs, long before we had our sophisticated hospital, was to participate in an operation on Oscar, one of our large and potentially lethal Orang utans. Oscar had a bad tooth and, not surprisingly, Jack Petty, our local dentist, was unhappy about examining the offending molar until the animal was well and truly asleep. In those days we did not know much about the mysteries of anaesthesia of these creatures and for everyone's peace of mind, and to keep our insurers happy, we had to press into service the local constabulary. A large bucolic sergeant was sent to us armed with a double-barrelled shotgun, in order to stand guard over Tony while he put Oscar under and Jack completed the dental operation. Tony said later that he was so thankful that Oscar showed no sign of coming round, for if he had he was quite sure that he would have received the benefit of both barrels of the gun and Oscar would have escaped unharmed.

'I don't trust Orangs,' Tony explained, 'but I trust large policemen with guns even less.'

Tony has now joined our ranks as General Administrator as well as veterinarian, and during a comparatively normal day has to deal with anything from a caesarean operation on a gorilla to a blocked pipe in the ladies' toilets.

Another facet of our work is in the field of public education, which is in addition to the professional training we provide at Les Noyers. Our Education Officer, Phillip Coffey, was once on our ape staff, looking after our Orang utans and gorillas. The patience, kindness and understanding required in caring for the great apes stood Phillip in good stead as he developed our education programme for schoolchildren. More than 7000 children use the Jersey Zoo as a 'classroom' every year, and there are 7000 more who belong to the 'Dodo Club', a society for our junior members. Phillip and David Waugh are putting the final touches to a 'Zoo Educators' Course' which will involve teachers from zoos everywhere, particularly those from developing countries. This is the first such course of its kind, and we are delighted that it is being held here and hope that it will be the forerunner of many more around the world.

We have a wide range of publications now which describe our diverse activities. There's our scientific journal, *Dodo*, of which Jeremy is editor-in-chief, and he is always cracking the whip over the heads of our staff, students and visiting researchers to get their papers in on time, whether they concern the techniques of hand rearing a rare fruit bat or the results of a long-term study on the ecology and behaviour of an endangered species in the wild. There is our newsletter, *On the Edge*, which goes out to all our members three time a year, and a special newsletter for the Dodo Club members, the *Dodo Dispatch*. Every issue of the *Dispatch* has a large colour poster of one of the species we're working with. We always do an extra print run of the posters, with a conservation message overprinted in the language of the animal's country, and these are sent out to schools, offices and shops in that country concerned to enhance local education programmes. To keep our former students of the International Training Centre in touch with us and with one another, we publish and circulate (now to nearly 300 people in 65 nations) a newsletter called 'Solitaire', which includes news about Jersey and our overseas projects, news from the students themselves about their work and notes on advances in conservation in these and other countries.

Of course, the staff do not spend all their time tending the animals, teaching courses or writing reports. In fact, they lead a fairly kaleidoscopic existence, for we send them out on field research trips and to supervise our *in situ* breeding projects. Thus Jeremy goes to Brazil periodically to keep an eye on our primate conservation work. David Jeggo goes out to advise on the breeding projects we have set up for the Caribbean parrot and to do censuses of wild parrot populations. Bryan Carroll (our Curator of Mammals) gathers information about fruit bats in the wild and Quentin Bloxam, our Curator of Reptiles, does similar work on reptiles in Madagascar and elsewhere. Sometimes the ways of obtaining the information we need are bizarre and would not occur to the average conservationist. For example, when we sent our Research Assistant, William Oliver, to do a field study on the Pigmy hog, he found there was only one way of finding out about their private lives. He caught a Pigmy hog and put a radio collar on it. Then, on elephant back, he would make his way through the tall grass and find out from his receiver how the tiny pig spent its days. Tracking the smallest pig in the world who is wearing a radio collar while you are perched on the back of an elephant is surely a rare experience for any conservationist.

So this is our multifaceted approach to species conservation, and it is no more nor less than what all other zoos in the world today should be doing to help save endangered species. But we have been greatly aided in our work over the years by developing two strengths for which I'm sure other conservation organizations envy us. The first is that we have managed to attract – and more importantly to keep – a wide variety of highly talented and dedicated people, and the second is that I am fortunate enough to write books which have become so popular that many doors have been opened to me which remain firmly closed to other organizations.

I think the secret of our fine staff is that we are not just 'another zoo'. We have laid down for ourselves certain goals and work towards them steadily, and this, I think, makes us unique and it is this that appeals to the people who join us. But on the whole we have been lucky in that staff tend to find us rather than the other way round, or at least they seem to swim into our net. The case of our Trust Secretary, Simon Hicks, is one very good example.

Simon, when we first met him, was director of that excellent organization the National Conservation Corps in the UK. People who feel deeply about aiding conservation can join this and give their time and skills for no financial reward, the reward being in cleaning out polluted rivers and village ponds, planting against erosion, building fences, dams and bridges and doing similar backbreaking but essential conservation work. Simon had brought a team over to help with some jobs we were doing. Tall, well built, with big blue eyes and curly red hair and a snub nose, thickly freckled, he arrived with his team and displayed an enthusiasm which was incredible. He vibrated energy in an almost tangible way. It was like standing next to a dynamo, not one of your paltry little dynamos but one of the sort they use on the *QE II*. He had enormous charm and handled his team with an efficiency I have rarely seen equalled. I was impressed. I consulted Jeremy and John Hartley. They had been impressed too. The work of the Trust had increased enormously and we urgently needed another man at the top to take over some of the load from Jeremy and John and it seemed that Simon was heaven-sent, if we could get him.

'Let's have another look at him,' I said. 'Let's get him over again on some spurious excuse: advising us on re-shaping the water meadow, or something.'

'Do you think that's necessary?' asked Jeremy, always cautious.

'Yes,' I said. 'Remember, Jeremy, the morning face on the pillow may look quite different from the one of the night before.'

Jeremy blushed. 'I see what you mean,' he said, doubtfully.

So Simon joined us and it was an infusion of the new blood we needed. His wild enthusiasm, his refusal to be beaten by any project suggested, however difficult or improbable, his sheer exuberance were boundless. It was neatly summed up by a South American friend. I said that she must meet Simon as it was an experience no woman should miss. I phoned down to his office.

'Si,' I said. 'I've a lady from South America here and I want you to meet her. Can you pop up for a moment?'

'Yes, yes,' said Simon, his voice clearly audible across the room where we sat. 'Jolly good, yes, be up in just one second.'

I put down the phone.

'Now listen,' I said.

Distantly one heard a bang, followed by what sounded like a rumble of thunder. Then there were two more bangs as from cannons of magnitude. This was followed by a thunder of feet on the stairs leading up to the front door of the flat that sounded as if we were being invaded by the entire Russian army, newly shod, followed by a bang on the front door that surpassed all previous bangs. My friend jumped and spilled her drink.

'Simon believes doors were designed to obstruct people who are trying to get things done,' I explained kindly.

He strode into the room, refulgent as a volcano, and crushed my friend's hand to a pulp. He talked and laughed and told jokes with immense charm for about ten minutes while he gulped a cold beer.

'Well,' he said, looking at his watch, 'must go, I'm afraid – got a crowd of volunteers to lick into shape – sorry I can't stay, see you again I hope. Jolly good. Splendid.'

He crushed her hand once more into his thumbscrew-like grip and left. She sank back on the sofa and listened to the rumble and bangs of his departure as one would listen to the retreat of an army.

'What did you say he was called?' she asked.

'Simon,' I said, 'Simon Hicks.'

'You should call him Hurricane Hicks,' she said firmly, and so Hurricane Hicks he has remained.

Simon's enthusiasm for his new job knew no bounds. Vibrating like a harp, in no time at all he had reorganized our animal adoption scheme so that, instead of just being a welcome but small contribution to the upkeep of various creatures, it now showed a handsome profit and we had a large waiting list of creatures for adoption. This scheme works on the principle that you adopt an animal, pay a contribution to its upkeep and get your name put up on the cage or enclosure. It seems to be particularly suitable for harassed parents or godparents who, as Christmas or birthdays roll around, cannot think of an original present, so they adopt an animal in the name of the youngster concerned. It works on a sliding scale, of course, so that a frog does not cost as much as a gorilla and, although the amount does not in any way represent the sum we expend on the creature's upkeep per annum, it is a most useful contribution towards it. It also gives both adults and children a sense of helping us in our work and a certain pride in coming to see 'their' animal.

When I originally started our sister organization in America, we had called it SAFE (Save Animals From Extinction), but we had to change it to the more pompous title of Wildlife Preservation Trust International since the Americans felt that SAFE sounded too like a prophylactic. Simon, however, had no such inhibitions and immediately patented the name and invented a new way for people to contribute to our work. He had cards printed for each species; each card had boxes labelled, for example, 'veterinary care', 'maintenance', 'return to the wild' and so on. People contributing could put a tick against the aspect of our work that most interested them and the money would be spent on that particular section of the work.

As well as these activities, Simon decided that he should go on the training course alongside our overseas students, and this he diligently did, cleaning and feeding the animals on every section in the zoo and, as he got more knowledgeable, taking groups of visitors around and telling them about our work and about the biology of the animals. He is, however, possessed of an enormous innocence, which makes him the perfect fall guy, so after listening to him talking to a group of Trust members (unbeknownst to him) I called him into my office. 'Simon,' I said, 'I want to talk to you very seriously.'

'Yes, what?' he said, a look of alarm on his face.

'It's these lectures you give to Trust members,' I said. 'I was listening to you the other day.'

'Oh, my God, you weren't, were you?'

'Yes,' I said severely, 'and while the bulk of your lecture was all right, I do feel you shouldn't have told all those poor people falsehoods.'

'Falsehoods?' he croaked.

'Well, maybe not falsehoods,' I conceded, 'I mean, you may really believe that Snow leopards come from the Sahara.'

He gazed at me and suddenly realized I was pulling his leg.

'My God, don't do that to me,' he said, 'You made my mouth go dry.'

It was at this period that Simon got into a complicated situation with our Orang utans. Our huge male Sumatran, Gambar, was extremely territorial and very possessive of his wife Gina. He now got it into his head that Simon – whose hair, roughly speaking, is Orang utan colour – was an undesirable male who had lewd designs on Gina. Whenever Simon approached, Gambar would hoist himself on to the bars and swing to and fro like a giant, ancient sporran, banging the hanging lorry tyre that swung from a chain in his cage to and fro with enormous thumping sounds, and then round off this terrific territorial display by grabbing Gina and, to Simon's acute embarrassment, mating furiously with her. It got to the point where Simon refused to go near the cage when taking Trust members round.

'Oh, look, Orang utans,' they would cry, espying Gambar from a distance.

'Yes, yes, jolly good,' Simon would say, feverishly. 'But I must just show you the lemur wood *first*.'

Simon confessed to me that Gina had such an accusing look in her eyes when Gambar was mounting her that it went straight to his heart.

'It's awful,' he said, 'it looks as if she's blaming me for it.'

'Never mind,' I said consolingly, 'just think how famous you'll be when Gambar makes his final move. The *News of the World* will pay you handsomely for your life story.'

'What?' said Simon. 'What final move? What are you talking about?'

'You'll be the only man in history who has been sued for alienation of affections by a Sumatran Orang utan,' I explained.

The popularity of my books has opened many doors for me, particularly among what are loosely classified as 'the famous'. It helps enormously if you know that the big boss of an organization is an avid fan of yours because you can then, quite unblushingly, phone him directly and ask him to do you a favour, instead of having to crawl slowly up the sticky and often obstructive bureaucratic ladder. Also, one person can lead you to a dozen others who may be of help, so the whole thing takes on the aspect of a daisy chain. The problem of finding a suitable patron for our then fledgling American sister organization was a case in point.

It was at about this time that Tom Lovejoy, having built up the American Board to a fine cluster of supportive and generous people, decided that what was now needed was a figurehead, someone well known to Americans and a person who would add lustre to the organization. Tom phoned me to discuss the matter.

'How about Prince Rainier?' I suggested. 'He likes animals and he's got his own zoo in Monaco.'

'His wife would be better,' said Tom shrewdly. 'No one in America has heard of Prince Rainier, but everyone's heard of Grace Kelly.'

'Very true,' I said, 'but I don't know her.'

'I bet David Niven does,' said Tom, 'and you know him. He was your gorilla's best man, wasn't he?'

'Yes, but I hate asking friends to trade on their friends, as it were.'

'He can only say no,' said Tom philosophically.

So I phoned David and asked his advice.

'I hate getting my friends involved in things like this, Gerry', he said, 'but I tell you what I'll do. I'll give you an introduction to her, but after that you're on your own, and I shall tell her so.'

'Marvellous, David,' I said, 'that's fine. All I want is the door opened and then Tom and I can do the rest. I'm sure it will be OK. Tom's got almost as much charm as you have.'

'Flattery will get you nowhere with me, Mr Durrell,' said David severely.

'The trouble with you second-rate actors is that you don't recognize the truth when you hear it,' I said, and rang off before he could reply.

In due course Princess Grace agreed to see us and a date was fixed. Tom was jubilant.

'Good ole little Miss America,' he said.

'She hasn't agreed to anything yet,' I warned him, and for God's sake don't go around calling her little ole Miss America. She's a princess, goddamit.'

'Only by marriage,' said Tom.

'You must learn, my dear Thomas, that the ones who become princesses by marriage are sometimes more conscious of their dignity and position than those who have been born to it.'

'Will I have to curtsey?' asked Tom.

'No,' I said, 'much as I would enjoy the sight of you curtseying to Princess Grace, I think it is something we can spare her. However, as you are so uncouth, I feel the least I can do is to give you a few lessons in how to behave when we meet her. I will give them to you when we meet in France.'

As it is not every day that you are invited to the palace at Monaco, I felt we ought to do the thing in style. Therefore, I and all my female entourage (my wife, my secretary and a long-time friend) were ensconced in an extremely lush hotel within note-rustling distance of the casino. Having jockeyed our tastebuds into a turmoil of expectancy with a delicate *kir* as an aperitif, we entered the dining room preceded by a most satisfactorily obsequious *maître*

d'hotel, and surrounded by a garland of attentive waiters. The delicious cucumber soup, cold as a Polar bear's nose, had been delicately sipped and the waiters had, in solemn silence, placed in front
of us the fresh salmon poached in champagne and cream, when
Thomas Lovejoy made his appearance. He looked like the sole sur-
vivor of one of the more unpleasant Mesopotamian earthquakes.
At the sight of him the *maître d'hotel* uttered a tremulous squeal
such as is wrenched from a tiny guinea pig trodden on
unexpectedly by a shire horse. I must say that I had a certain
sympathy with the man.

In one hand Tom clasped what seemed to be all his worldly
goods incarcerated in a briefcase which had apparently been constructed out of the skin of an ancient crocodile suffering from lep-
rosy. His suit looked as though it had been slept in by seventeen
tramps and then discarded as being of no further service. His shirt
was fish belly grey, except for the area around his neck, which was
black. His tie – at one time I have no doubt a magnificent piece of
neckwear – looked as though it has been seized and thoughtfully
masticated by one of the less intelligent dinosaurs and then regur-
gitated. His shoes completed the whole ensemble: Charles Chaplin
had spent years trying to get his shoes to look like that without
success. Carunculated and furrowed as any chestnut, the toes
standing up like flagpoles, the soles in imminent danger of losing
their grip on the upper part of the footwear, they were shoes in
which you felt – should you be so unwise as to investigate them
closely – might lurk any number of communicable diseases.

'Well, hi there,' said Tom, packing his unsavoury body into a
chair. 'Sorry I'm late.'

My female entourage regarded him as if he were a toad found
lurking in their soup. There they were, clad in silks and satins and
raiment fine, delicately made up, exuding expensive scent like a
newly-mown summer hay field and into their midst had lurched
the Phantom of the Opera.

'We hope you're not going to meet Princess Grace at the
palace looking like *that*,' they said ominously and simultaneously.

'Why?' asked Tom, puzzled. 'What's wrong with me?'

They told him. In a long life of listening to women dissect their menfolk, I have never heard anything more comprehensive and derogatory. He was rushed upstairs and, regardless of his protests, was stripped of every article of wearing apparel he possessed and, while he sat on my bed wrapped in a towel, the cleaning brigade went into action.

'I don't see what's the matter with the way I look,' he said aggrievedly, 'I was having lunch with the President of Peru yesterday and he didn't say anything.'

'There are some women who judge a man by his clothing, and I am sure that Princess Grace is one of those. They have special eyesight and scanning equipment. Take them into a room full of three hundred people and they will immediately detect a microscopic spot of egg yolk on the tie of a man standing at the far end.'

'I can't help it if my luggage went astray in Paris, can I?' he said.

'Well, you can't tell her that,' I pointed out, 'she'll think you go around all the time looking like a rag-and-bone man.'

Presently, Tom's wardrobe reappeared, looking slightly less like a care package from the Crimean war, and as he dressed I gave him some lessons in court etiquette.

'Remember to bow when you shake her hand,' I said, 'and only call her Your Serene Highness or Marm.'

'Marm what?' asked Tom.

Marm, you know, an abbreviation of Madame.'

'You mean like a schoolmarm?'

'Can't I call her Mrs Rainier?'

'No, you certainly can't. Just stick to Your Serene Highness.'

So we got in a taxi and went up the hill to where the pink fairytale palace stood looking out over the glittering Mediterranean. We were stopped at the gate by an efficient-looking sentry wearing a uniform which might have been designed for *No, No, Nanette* or some similar musical extravaganza set in Ruritania. We were taken to the abode of the princess's secretary, where we were told to wait for a moment.

'Classy bit of real estate,' said Tom, looking around at the marble and gold leaf.

'Now, for heaven's sake, remember what I've told you,' I said. I felt that as Tom was the chairman of our American Board he should meet the princess first. Presently, the door was opened by the secretary and we were ushered into the private office. Dazzlingly beautiful and elegant, Princess Grace rose from behind her desk and came forward, smiling, to greet us. It was then to my horror I saw Tom wave a friendly hand at her.

'Well, hi there, Grace,' he said.

Trying desperately to undo whatever damage had been done, I spoke up manfully.

'Your Serene Highness is most kind to spare us the time,' I croaked. 'This is Doctor Lovejoy, chairman of our American Board, and my name is Durrell.'

'Do come and sit down and tell me what you want,' she said, smiling the sort of smile that turns any red-blooded male into a babbling idiot.

So Tom and I sat on a large sofa, Princess Grace between us, and tried to explain the purposes of the Trust. Although the princess listened patiently, I got the very strong feeling that the answer would be no. I felt she had seen us only because of her friendship with Niven and she was now searching for a polite way of refusing our request. So I played my trump card. Just as she was saying she had so many other commitments, and that she did not really think . . . , I slid on to her lap a large photograph of our newly-born baby gorilla, lying on its tummy on a white terry towel. The words died away on her lips and she uttered a schoolgirl-like squeak of delight. She did not actually say 'Diddums', but you felt that with very little encouragement she would.

'Your Serene Highness, these are the sorts of animals we are trying to help,' I said.

'Oh, it's so cute,' she cooed, 'I've never seen anything so cute. Please can I show it to my husband?'

'It's yours, I brought it for you,' I said.

'Oh, thank you so much,' she said, her eyes still fixed on the photograph, misted with love. 'Now tell me how I can help.'

Ten minutes later we left the palace with Her Serene Highness, Princess Grace of Monaco, as our American Trust's patron.

'I knew that gorilla picture would get her,' I said to Tom jubilantly as we climbed into our taxi. 'Every woman I've shown it to has gone nuts about it. It brings out the mummy in them.'

'I don't think it had anything to do with the picture,' said Tom.

I stared at him.

'What d'you mean, it had nothing to do with the picture?' I said. 'It was the picture that clinched it.'

'No, what really won her over was the little piece of egg yolk I had on my tie,' said Thomas, grinning.

5

Return to the Wild

A good many years ago, when we had just started the Trust, I would try to point out to people the point and purpose of captive breeding. Their inevitable question was 'What have you put back?' as if the whole exercise consisted merely in breeding a few specimens, bundling them into crates, shipping them back to their country of origin and flinging them out into the nearest bit of forest. Nothing could be further from the truth.

The whole business of captive breeding for conservation is one beset with problems, but once overcome, i.e. getting successfully through stages one and two of our multifaceted approach, you can begin in earnest on stage three, which is to put captive-bred stock back in the wild, in places where the species has become extinct, in new areas within or near the species' natural range which have suitable habitat, or in areas where an endemic wild population needs an infusion of new animals. Stage three is the trickiest part of all.

The tricky thing about returning captive-bred animals to the wild is that it is a wholly new concept, a wholly new art if you like, and we are learning as we go. To begin with, no two species are alike in their demands, and the wants of each have to be learnt as a vital preliminary. Second, you cannot take an animal which may be the third or fourth generation born in captivity and simply push it back into the wild. Surrounded by food it would in all probability perish, for it would be used to having its fruit or whatever cut up and served in bowls. It would be the same as taking a millionaire of long standing out of the Ritz and making him sleep on a park bench covered with newspapers and forage for his food in dust-bins. It would take time to indoctrinate him.

The methods so far evolved are fairly straightforward but, as I say, the whole process has to be adapted to the individual animal and, for that matter, to the individual place. Our first attempt at returning the Pink pigeon to the wild in Mauritius was an example of how easily things can go wrong. We had decided to do our first release of the birds into a sort of 'halfway house' in the botanical garden of Pamplemousses. In this vast area there was a plentiful supply of leaves and fruit and, because there were plenty of access roads criss-crossing the gardens, it would be possible to monitor closely the birds' movements and reactions. So we constructed a special release aviary, one side of which would hold two pigeons to be released and the other two more which we hoped would serve as 'decoys' to encourage the freed birds to stay in the gardens.

The first birds were carefully chosen from the breeding centre at Black River, the complex of aviaries and enclosures built and staffed by the Mauritian government and operationally funded in great part by our Trust. Once they had settled down and adapted to their new home the great day of the release came. I was in Mauritius for this auspicious occasion and I was to release the birds. The idea was that once the birds were freed – or at least had access to freedom – they could still use the aviaries as sanctuary and a constant food supply would be kept there until such time as they decided that they were self-sufficient. I went along on the appointed day and with a great flourish – for it was an important occasion – I pulled the string to raise the flaps which would allow the pigeons their freedom.

The string broke.

There was an embarrassing pause while someone was dispatched for more string. At that moment I sympathized deeply with those ladies, dressed in their best, who keep banging bottles of champagne against the bosom of new ocean liners without effect. Eventually, string was procured and the flaps duly raised. The pigeons behaved beautifully and flew out and sat on top of the aviaries. We then expected them to take advantage of their freedom and fly off into the trees. Not a bit of it. They sat stolidly on top of the aviaries without blinking an eye, looking like mentally retarded examples of an amateur taxidermist's work. How I wish all the idiotic people who prate about the cruelty of captivity and the joys of freedom could have seen those birds.

It was unfortunate – in hindsight – that the authorities wanted this, the first release, to receive no publicity, and at the time it seemed a sensible enough request. We released eleven birds altogether into Pamplemousses, and they eventually plucked up courage and began to sample their new-found freedom by investigating all areas of the extensive gardens. They soon had a new hazard to face. Small boys with slingshots patrolled the grounds endeavouring to get specimens of the Common dove for their tea. They could not be expected to tell the difference between a Pink pigeon and a dove, except that one was fatter and therefore more desirable (in fact, the flesh of the Pink pigeon has an unpleasant taste and is therefore inedible). This, combined with the Pink pigeon's trusting – almost imbecile – nature was fatal. Several of the pigeons fell to the young hunters' deadly weapons. One wonders, as I say, whether a big publicity campaign would have made any difference. Are small boys with slingshots affected by publicity campaigns? No one can be certain. But in spite of the small boys, some of the birds paired up and produced babies, although they deserted them at the nestling stage, probably as a result of human disturbance.

Suffice it to say that this our first release was not a resounding success, and so we caught the birds and took them back to Black River. Still, we had learnt a lot from this 'practice' release. The birds did not immediately fly off into the wild blue yonder, but stayed near the release site, where they could be fed until they got used to their surroundings. Then they began feeding on the leaves and fruits of exotic as well as native plants, an important observation because exotic plants have invaded the Mauritian forests to which the birds would one day be returned. Finally, we had proved that captive-bred birds could and would reproduce outside of captivity.

So with great confidence we planned the next step: a real release to the wild in a remote part of Mauritius known as the Macchabe Forest. Again we built aviaries and established the birds in them. This time, before release, some of the pigeons were fitted with tiny radio transmitters, so that we could track them, for it was one thing to find a bird in the botanical gardens and quite another to find it in the thick forests and deep valleys of Macchabe. A few days after release, the birds followed no set pattern. Some flew off beyond the range of their transmitters and disappeared for weeks at a time, only to reappear mysteriously. Others flew into the forest but reappeared at their aviaries every day for supplementary feeding. Others flew only a few hundred feet from the aviaries and remained there for months. Gradually, it became apparent that the birds were getting acclimatized to the wild state and were becoming daily more self-sufficient, choosing the leaves and fruit that took their fancy. Nevertheless, we kept up the supplementary feed as a precaution, for we did not want a wild source of seasonal food to disappear suddenly and the birds to starve in consequence. At the time of writing, two of the released pigeons have appeared back at the feeding platform each with a youngster in tow. So far we can say, cautiously, that this new release has been a success.

This is excellent for two reasons. It now proves that birds several generations born in captivity can re-adapt to the wild state, and, probably more importantly, we have established the bird outside what we have always called the Pigeon Wood. This is a valley set deep in the mountain forests, consisting of a small stand of exotic cryptomeria trees. The entire wild population of pigeons (probably less than twenty-five in 1978) nest in these. To have all the birds nesting in an area covering a few acres was, of course, dangerous in the extreme, for a happy band of the introduced monkeys or an equally happy primate – a man with a shotgun – or a really bad cyclone could have wiped out the Pink pigeon forever. Yet stubbornly the wild birds would nest nowhere else. It was a case of having all your eggs in one basket with a vengeance, but, by establishing a new colony in Macchabe we hope that the birds will get fixated on this new area and thus found a new colony. Having done this, we can continue to found other colonies in suitable patches of habitat throughout Mauritius so that should anything untoward happen to the original wild colony we have not lost the species.

Quite another problem has presented itself with the Waldrapp or Bare-faced ibis, because it insisted on nesting in a place where a town has spread. These are large birds, clad in dark plumage that becomes iridescent when the sun catches it. They have bare faces and long dark coral red beaks, and habitually wear a slightly affronted expression. Their vocal repertoire is large and comic, consisting of a series of choughs, whirrs, growls and the sort of noise that precedes a really good expectoration in Latin countries. At one time, the distribution of this extraordinary and most useful bird was wide, ranging from Turkey and the Middle East along North Africa and into practically the whole of Europe, nesting as far north as the Alps. In mediaeval times the young were considered a delicacy to be eaten only by noblemen, though I would not be surprised if some of the plump babies found their way into the poorer peasants' cooking pots. The earliest written references to this extraordinary bird date from the sixteenth century and emanated mainly from the town of Salzburg, where it was described as the Forest raven. There were some protective laws passed by King Ferdinand and the Archbishops of Salzburg in 1528 (presumably so that the aristocrats could go on eating the young birds and the peasantry could not) but these laws proved useless.

When I say the bird was useful, it – like so many birds – was a natural pest control agent, feeding on the larvae of noxious insects, as well as frogs, small fish and small mammals. When you consider the bulk of the bird and the fact that it has up to four young, a considerable quantity of insect larvae must be devoured to keep these birds going. At one time the arrival of the birds at their immemorial nesting sites on cliffs heralded the arrival of spring and so, particularly in the little Turkish town of Biriçek, the birds' return was a signal for a festival. But then that unpleasant substance DDT was invented and used, as it has always been and still is, indiscriminately. In consequence the ibis began to suffer, for their food supply was contaminated. They had already been hunted out of their ancient nesting areas in Europe, and in the Middle East and North Africa their numbers were dwindling rapidly. In fact, the colony at Biriçek was the only eastern one, but the town had grown apace. Many of the new buildings which had sprung up around the cliff where the ibis nested had flat roofs on which, because of the intolerable summer heat, the inhabitants used to sleep. Needless to say, they did not take kindly to the fact that the birds (who, after all, had been there first) would deposit a considerable amount of guano on the sleeping inhabitants of the town. So the birds were stoned and shot at, and what used to be an occasion for a festival became a nuisance. The Turkish government tried in vain to protect the birds, but under the pressure of human attack on the one hand and insecticides on the other the last eastern population started to falter and fail. At the time of writing, there are *no* wild birds left in the east. The only other known wild populations are small vulnerable ones in Morocco, Algeria and Saudi Arabia.

It is fortuitous that the ibis were established in captivity in both Innsbruck and Basle Zoos, and it was from the latter that we got the founder members of our flourishing colony. We have satellite groups at Edinburgh, Chester and Philadelphia Zoos and we have funded the building of aviaries in Morocco for captive-bred stock from us and from other European zoological collections. The young bred in the aviaries will form the nucleus of a release programme in carefully selected sites. We are already making plans for similar reintroductions in other parts of North Africa. According to Egyptologists, it is possible that it was the Bare-faced ibis that was the first bird Noah released from the Ark.

Thus it would be a heartwarming idea to establish a wild colony of these birds on a cliff face near Luxor, for example, and have them flying among the vast ancient monuments as they did when those same monuments were being hewn out of massive rocks.

We have yet to try a release with the ibis, but a few years ago, just after we had begun the pigeon release, we were ready to go ahead with another species, this time a mammal, a creature called the Jamaican hutia. This had all the hallmarks of success, but what happened shows how a project which on the surface seems simple and straightforward can develop unsuspected pitfalls if you are unwary.

Hutias are a group of rodents confined to the Caribbean Islands. There are different species in the Bahamas, Cuba and Jamaica. The Jamaican species, locally called a 'coney', is a browny-green animal, about the size of a miniature poodle and looks not unlike an enlarged guinea pig. They are the only large indigenous surviving mammal found on the island, although at one time they were abundant and provided a major food source for the original inhabitants as well as for the indigenous Jamaican boa constrictor. However, excessive hunting with modern weapons and destruction of the forests in which they live put them in peril. The Trust received its first hutias in 1972 – two males and a female captured in the John Crow Mountains – through the good offices of a Trust member, and eight more were acquired in 1975. From these came the first captive birth ever recorded, and during the next 10 years 61 litters comprising 95 young were produced. Of these, acting on our principle of never having all your eggs in one basket, 19 were sent on breeding loan to 6 other collections in 4 different countries.

Back in 1972, just as our splendid new hutia accommodation was nearing completion, I got a telephone call from Fleur Cowles, one of our trustees. She told me that the Hollywood star, Jimmy Stewart, and his wife Gloria were going to visit her and that she was going to bring them over to Jersey. Always with an eye to the main chance, I asked if Mr Stewart would like to open our new hutia breeding unit to provide some publicity for the Trust. Back came the answer that he would be delighted.

On the appointed day I went down to the airport to meet them. Stewart was unassumingly himself, walking with a slight cowboy slouch, drawling sentences in his lovely husky voice. Gloria was a handsome woman, immaculately groomed as only a wealthy American can be, with immense charm but a slight glitter in her eyes which told me she could easily resemble one of Mr Wodehouse's famous aunts if things did not turn out to her satisfaction. She was the sort of spirited person to whom *maîtres d'hotel* give instant allegiance and servility, in case worse than their wildest nightmares should ensue. As we waited outside the airport for John to bring the car round, talking about this and that, James Stewart suddenly disappeared. One minute he was there – tall, gangling, a gentle smile on his face – the next he had softly and silently vanished like a puff of smoke. One would have thought it impossible for such a big man (in every sense of the word) to eclipse himself so deftly without anyone noticing.

'Where is Jimmy?', Gloria asked suddenly and accusingly, as if we were concealing him from her. We all looked around vacantly.

'Perhaps he has gone to the comfort station,' I said, using an American euphemism I adore.

'He did that on the plane,' said Gloria. 'Where on earth is he?'

Having eliminated the comfort station as a possible hiding place, I could not think for the life of me where he could have gone. Gloria's increasing agitation infected me with a sense of unease. Had he been kidnapped? I could see the world headlines in the illiterate press: 'James Stewart snatched at hutia party – famous actor becomes as extinct as the animals he went to visit.' This was not the sort of publicity I was seeking for the Trust.

At that moment, John rolled up in the car. 'Shall I go and tell Mr Stewart the car's here?' he asked.

'Where is he?' everyone asked in unison.

'He's out there on the tarmac looking at a plane,' said John.

'Go and get him, please,' said Gloria. 'He can't keep his hands off planes.'

'How did he get out there?' I asked, for airport security is very tight in Jersey.

'Can you imagine anyone stopping him, seeing who he is?' asked John.

Presently, the truant loped back into our midst.

'Er . . . kinda nice little plane out there,' he explained. 'Yeah, sorta little job, very neat. Kinda cosy, you know. Neat. Hadn't seen one before.'

'Get into the car, Jimmy,' said Gloria, 'you're holding everyone up.'

'Yeah, yeah,' said Jimmy, either unrepentant or not listening. 'I'm glad I saw that. Kinda neat.'

After we had lunched he opened our hutia nursery with great charm, saying that he had always liked Hoot Ears ever since he first met them, which was about five minutes ago. This ordeal over, we took them out to dinner at a friend's house.

Over drinks in the conservatory and the excellent meal that followed it, Jimmy seemed preoccupied. I think he was suffering from jetlag, which has a stultifying effect on anyone. The meal over, we repaired to the drawing room where Jimmy carefully lowered his gangling shape into the bosom of an enormous sofa. His eyes wandered vaguely round the room and suddenly focused on something that interested him.

'Gee, it's a piana,' he said, his eyes fixed longingly on the baby grand that crouched in the corner.

'Jimmy, no,' said Gloria Stewart, warningly.

'Yes sir, a piana,' said Jimmy, with the delight of one making the discovery of the century, 'a kinda little baby piana.'

'Jimmy, you're not to,' said Gloria.

'A little toon . . .' said Stewart musingly, starting to unravel his length from the sofa, a fanatical gleam in his eye, 'a toon – what's that toon I like?'

'Please Jimmy, don't play the piano,' said Gloria desperately.

'Oh, I know . . . "Ragtime Cowboy Joe" . . .,' said Jimmy approaching the instrument, 'Yes siree, "Ragtime Cowboy Joe".'

'Jimmy, I beseech you,' said Gloria, her voice breaking.

'Yes, a kinda nice, swinging toon, that.' Jimmy seated himself on the piano stool. He lifted the lid and the baby grand grinned at him like a crocodile.

'Now – er – let's see – er, how did it go,' said Jimmy, plonking his long fingers on the keys. We were immediately apprised of two facts. The first was that Jimmy Stewart was tone deaf and the other that he could not play the piano. In addition, he had forgotten all the lyrics except the basic one of the title. In all the years I had watched his impeccable performances on the screen, I had never seen him do anything like this. He played all the wrong notes and sang out of tune, trying to make the two match. In his husky, croaking voice he sang the title of the song over and over again, going back to the beginning when he thought he had missed something out. It was like watching an armless man try to swim the English Channel and yet it was excruciatingly funny, but you did not dare laugh as he was taking such pride in his performance. In the end, he exterminated 'Ragtime Cowboy Joe' to his satisfaction and then turned to us, happy in his achievement.

'Would anyone like to hear some other toons?' he enquired generously. I was tempted to ask for the 'Star Spangled Banner', but it was not to be.

'Jimmy, we must go.' said Gloria.

And go they did.

To have been given a performance like this by the great James Stewart was an honour, but I was sure his wife did not agree.

It is always exciting when a new animal arrives at the Trust's collection, is released into its new quarters and you can watch it settle down. The Hoot Ears, as christened by Jimmy Stewart, however, proved to be the exception rather than the rule. Handsome, portly animals though they were, with heavy hindquarters which made them look as if they were wearing trousers several sizes too large for them, they lacked the scintillating personality one might have hoped for. They displayed all the *joie de vivre* of a bevy of church-wardens attending the funeral of one of their number. There was only one thing they did which could possibly be described as eccentric. Like most creatures, they had not read the textbook description of their behaviour and so they did not know that they were supposed to be strictly terrestrial. Ponderously, and with total lack of expression, they would climb up the branches in their cages and perch near the ceiling – one supposed imagining themselves to be flocks of flightless birds. True, I did see the young ones frequently indulge in what could be described as 'catch-as-catch-can' games, but they were of a very staid variety and one was reminded of over-weight Victorian children indulging because their elders expected it of them.

When the numbers of young we had bred were sufficiently high, we started thinking in terms of reintroduction. Our then Research Assistant, William Oliver, went out to Jamaica to fix up all the preliminaries, which included the selection of a suitable site (a place that seemed satisfactory from the hutias' point of view, particularly freedom from hunting pressure) and the involvement of Hope Zoo in Kingston in the venture. A total of forty-four of our Jersy-bred hutias were sent out in 1985–6, and settled in their family groups in specially built cages at the Hope Zoo. Meanwhile, an extensive vegetation survey was done on the chosen site to make sure that the hutias would lack for nothing in terms of food-stuffs. Then they were transferred to the release site, each family group into a temporary enclosure surrounding a specially con-structed, semi-artificial rock warren or 'coney' hole. After a week or two, when the animals seemed to be used to their new situation, the fence was removed and the progress of each group was moni-tored for up to three months.

Early reports were most encouraging. Only three animals dis-appeared during this initial monitoring period, but the rest of them rapidly became self-sufficient and remained in good condition. Our hopes were high that the reintroduction was going to be a great success. However, when the site was reinvestigated later in the year only eight hutias could be located. These animals, which included two conceived and born on the site, were all in excellent condition. However, no others were found during a six-week search. In the following year only two animals were found, one a Jersey-bred specimen and the other thought to be wild born. Both were in good condition, but the whereabouts of the rest of the specimens was, and remains, a mystery. The animals that had been released had, early on, adapted to the wild excellently and behaved as normal wild hutias do. This site seemed eminently suitable with a plentiful food supply and freedom from hunting pressure. We had to conclude, therefore, that the disappearance was due to illness or to predation by feral dogs and cats. However, we haven't given up hope – literally – because we are now working with the Hope Zoo to establish a sufficiently large breeding colony there from which a second reintroduction, with the help of students from the University of the West Indies, will be attempted.

All the frustrations involved in releasing animals to the wild are more than made up for when you join forces with people and meet with success, as in the case of the Golden lion tamarins. These enchanting creatures, smallest of the primates, along with their close relatives, the marmosets, live in the coastal rainforest of Brazil. Unfortunately, this special rainforest has been ruthlessly and thoughtlessly destroyed and all that is left are pockets of trees, some not even connected with one another, so that the animals of each of these pockets are isolated and cannot renew their species' genetic resources by mixing and mating with others of their kind, even if they are only a few miles away. At one time, the Atlantic coastal rainforest covered an area of 135,000 square miles. Now less than five per cent remains and this is being steadily whittled down by axe, fire and bulldozer. As this forest is stripped, it not only drives to extinction – or its brink – the tamarins, but the myriad other creatures and plants that go to make up this extraordinary ecosystem. When you fell a tropical tree you are doing the equivalent of destroying a huge city, because of the thousands of creatures that live in, on and around it.

The Golden lion tamarin is probably one of the most beautiful of all mammals. A little bigger than a newly-born kitten, it has incredibly long 'artistic' fingers and its long fur looks, quite literally, as if it is spun gold. This amazing glittering pelt stands away from its face in a sort of semi-recumbent mane which gives it a lion-like look. Like all the marmosets and tamarins, their movements are incredibly quick and sometimes they move with such speed it is impossible to follow the movement with your eye. They are omnivorous, the bulk of their food being fruit and insects, but they will eat tree frogs with relish and will even (it has recently been discovered) go into hollow trees in the daytime to hunt for roosting bats to add to their diet. Their vocalizations are very birdlike as they communicate in a series of trills, sharp squeaks and chatterings.

In addition to the destruction of the forest, these beautiful little animals had been popular with the pet trade and for biomedical research, so by the late sixties and early seventies it was apparent that the species was in serious danger. It was estimated then that no more than 150 individuals were still living in the fragmented forest blocks which remained. This alarming state of affairs was highlighted by the brilliant work carried out by Dr Adelmar F. Coimbra-Filho, now director of the Rio de Janeiro Primate Centre. In 1972 a conference was held, during which the plight of these animals was discussed and an attempt made to assess both wild and captive populations. It was obviously of the greatest importance that self-sustaining captive populations were established while, at the same time, trying to address the problem in the wild. It is mainly due to the dedicated work of Dr Devra Kleiman that this has been so successful. Between 1972 and 1980 very few zoos had Golden lions and these were mostly American. These zoos carefully expanded their small populations, and the result was spectacular. The captive population sprang from 153 to 330 – about double the wild population – within five years. Fifty to sixty

Golden lions were being born every year and so there was now a sufficiently large and stable population to start thinking about putting some captive-bred specimens back into the wild. The success of this project was due to the formation of a consortium of zoos for the management of this species.

In 1978 we received our first pair of Golden lions and also joined the consortium. The arrival of our Golden lions caused quite a sensation. It is one thing to see a painting or colour photograph of a creature, quite another to see the animal in the flesh. These tiny primates, glittering like doubloons, raced about their cage at such speed they looked like ingots being thrown about. As they whisked about exploring their new domain, they kept up a chorus of chirrups, squeaks and chitters as if each were a miniature tour guide telling the other where it was and what to look at.

Finally, when they had settled down, they became the centre of attraction in our marmoset range for they were by far the most striking and attractive of this enchanting group of primates. Finally, came the day when the female successfully gave birth to twins (the normal complement), two minuscule little gold nuggets which could each easily have fitted into a small coffee cup. At first, clinging to the dense fur of their parents and matching it so beautifully, they were extraordinarily difficult to see, for their little faces were smaller than a fifty-pence piece. As they grew older they grew bolder and would leave the security of their parent's body to explore the cage on their own, though always ready to fly back to the security of the parent's fur at any imagined danger. To see them in the sunshine chasing butterflies unwary enough to drift through the wire mesh was an entrancing sight. Not only was it an incredible, dainty ballet as they twisted and turned, leapt and scuttled after the pirouetting insects, but as the light caught them their coats sparkled in myriad colours from sandstone red to the colour of the palest wedding ring. For some reason my suggestion that the babies be christened Fort and Knox respectively met with such antagonism from all quarters that, outnumbered, I was forced to relinquish the idea.

Meanwhile the plans for release into the wild of captive-bred lions were moving ahead. Naturally, a plan of this magnitude had to be approached with great caution and attention to detail. An ecological survey had to be done to assess the wild population of Golden lions and, this done, to locate an area of forest uninhabited by a wild population but suitable for the release of the captive-bred specimens. Meanwhile, fifteen animals from five US zoos were chosen and sent to the Rio Primate Centre for training. An animal which is perhaps the third generation born in captivity is used to set mealtimes and never has to go out and search for its food. Most important of all in the cushioned world of captivity, there are no predators in the shape of snakes and hawks, and even *Homo sapiens* is considered an obliging gift-giving friend. So the animals have to be introduced slowly to the stern realities of life in the forest if they are to survive. At one point it was discovered that they were alarmed and daunted by tree branches which bent. In the well-conducted zoos they came from the branches were rigidly nailed into place, so a branch which gave under your weight was an alarming experience until you learnt how to cope with it. They had to learn how to incorporate into their diets wild fruit they had never seen before and here it was discovered, fascinatingly enough, that the younger animals were quicker at learning this and were showing the older ones what to do.

The initial releases got off to a slow start, but as the animals *and* the people in charge of the project learnt more and more they were finally successful. One photograph shows a captive-bred specimen eating a frog, an item never included in her diet in Washington, and proof that the animals had settled down in their environment. The next phase involved releasing captive-bred animals with wild ones, and it was a great day when twins were produced by a female born in captivity but who had mated with a male born in the wild. By this time we had bred over twenty-five Golden lions in Jersey and so were able to take part in the venture by donating five of our animals. These were released as a family group in a patch of forest with no wild tamarins present, and we're

very proud to say that our group was the first in the project to produce offspring from parents which had *both* been born and raised in
captivity. This is proof, if proof were needed, that if all the various
disciplines involved work in harmony towards a common goal,
captive breeding can and does work, and with it we should be able
to pull back innumerable species from the brink of extinction.

I always remember having a delightful picnic lunch with Roger
Payne and his family on my second visit to America. It is Roger, of
course, who has done so much wonderful whale research and is
responsible for those mournfully beautiful whale songs to which
one listens enraptured, longing to know what these huge and
extraordinary animals are saying to each other. However, during
the course of the picnic, Roger asked me what the Trust was all
about and I endeavoured to explain our aims and objectives.

Finally, Roger said, 'I think I see what you mean – you're
breeding them to put back *there*, providing there is a *there* to put
them back into.' Thus, in one pithy sentence, he highlighted one of
captive breeding's great problems: call it the 'There Syndrome' for
want of a better description. The situation with the Rodrigues
fruit bat is a case in point.

Rodrigues is a small island which lies 406 miles east of Mauritius. It
was described by an early settler, Leguat, as a paradise, thickly
wooded and overflowing with wonderful creatures. There was the
Solitaire, a strange long-legged ground bird not unlike today's
African Secretary bird. There was a species of giant tortoise found
in such profusion that it was said you could walk a league on their
backs without touching the ground, rather as a squirrel in olden
times in England could go from London to Aberdeen without
descending to earth, using hedges and woodland as a highway. Also
on this tropical paradise was a gecko three feet long, a ground
parrot and many other wonders. Now Rodrigues lies in the blue
sea, under a fierce sun – dry, eroded, desiccated – with scarcely any
greenery left and a burgeoning human population. Gone are the

long-legged Solitaires, gone the ground parrot and the giant gecko, gone the cobbled streets of tortoise backs. All that remains are a few tiny patches of forest, and in the largest, called the Cascade des Pigeons, lives a colony of golden-furred fruit bats, found nowhere else in the world. John Hartley and I had gone to Rodrigues in 1976 and collected a breeding nucleus of eighteen bats from the small colony of 120. Some were established at the Mauritius government's aviaries at Black River (where the Pink pigeon is also kept) and three males and seven females were brought to Jersey. From these we have reared ninety babies and have established satellite colonies in both Great Britain and America. We have now bred enough to think of reintroduction. But where to?

It is obvious that Rodrigues cannot support them. The colony from which John and I collected the breeding groups has now risen, thankfully, to about 800 animals, as there have been no major hurricanes recently and human disturbance in Cascade des Pigeons has been minimized. But 800 bats are thought to be about as many as the remains of Rodrigues' natural forests can cope with. Even if the current reafforestation programme is a success, it will be many years before the new forests could support bats, and even if it were possible to set up another colony in Cascade des Pigeons the problem of hurricanes remains. It is only a matter of time before an enormous one hits this tiny island and removes the trees of Cascade des Pigeons as the wind whips a scarf from one's throat.

We had thought of the Chagos Archipelago lying just over a thousand miles north of Rodrigues. These islands are uninhabited and outside the cyclone belt and the trade winds. It is presumably because of this that no bats flew or were blown there to colonize them. There are three atolls in the group which could conceivably support a fruit bat colony: Diego Garcia, Peros Banhos and Salomon. At one time the islands were planted with copra but the plantations were abandoned in 1972. Some of the fruit trees and vegetable gardens established by the copra growers are going wild, and these might provide a food source for any introduced bat.

Needless to say, like so many pleasant islands Diego Garcia is a military base now, and access is strictly controlled. The other two islands, however, may possibly be able to sustain a bat colony, although it is always unwise to introduce alien creatures into an ecosystem. One has seen what havoc they produce – be it deer in New Zealand, rabbits in Australia or donkeys on the Galapagos. However, in the case of these islands, all the natural vegetation was removed to make way for copra. In addition, rats, cats, pigs and goats were introduced and have become feral, with all the concomitant effects that such creatures have on any ecosystem. Therefore, the introduction of the Rodrigues fruit bat could not worsen the situation, but could help save the species from extinction and possibly be beneficial to the regeneration of the vegetation as well.

It has been proved that fruit bats play a very important 'gardening' role in forests as flower pollinators. Moreover, they eat the fruit, the seeds pass through their bodies as they fly from place to place, drop to the forest floor and take root, spreading that particular species of fruit tree across a wide area. It is never wise to say – as so many people do in scornful tones – what use is it? of any creature or plant, for it assuredly has a use, though it may not be immediately apparent. The horticultural activities of bats aid a host of other life forms, including man himself.

We asked permission to visit Diego Garcia and permission was refused, the authorities telling us to focus our attention on the other two islands. This we are doing in the hopes of finding a safe haven for our little flying golden teddy bears, for that is what they look like. We hope that by the time we find this haven it will not have been turned into an atomic bomb range. Anything is possible in a world where killing seems to be more important than preserving.

While the Rodrigues fruit bat is suffering from the 'There Syndrome' , some other animals from the Indian Ocean are being cured of it. This is the story of Round Island, which I think is our most impressive achievement to date: we have rescued an island and its unique inhabitants from oblivion.

*   *   *

When John Hartley and I were in Mauritius setting up the breeding project for the Pink pigeon, my attention was drawn to the
problems facing Round Island, a volcanic cone some 350 acres in
extent, lying thirteen miles north-east of Mauritius. The extraordinary thing about it is that on this small scrap of earth live no less
than two lizard and two snake species and several plant species
found nowhere else in the world. Furthermore, it is one of the very
few elevated tropical islands in the world free of rats and mice, and
it is an important breeding station for various seabirds. Many years
ago Round Island resembled Mauritius in miniature: that is to say
the high parts of the island were thickly forested with hardwood
trees including ebony, while on the lower slopes lay an apron of
palm savannah. Then sometime in the 1800s some fool released
some goats and rabbits on the island, of all animals the most
destructive. The result was like shutting a Sabre-toothed tiger in a
sheepfold. What the goats did not eat the rabbits did, and soon the
hardwood forest had disappeared altogether, the palm savannah
was fighting a rearguard action and the rapidly eroding island was
slowly but surely starting to slip into the sea. On my first visit it
was looking like the raddled, seamed death mask of a centenarian
Red Indian, with only a scattering of palms, a few pandanus and
some scant low growth left. It was obvious that something had to
be done very quickly about the reptiles, for their habitat was being
munched away as it grew. John and I paid two visits to the island,
for the Mauritius government was in complete agreement that a
nucleus of the reptiles should be captured for a captive-breeding
initiative first in Jersey and later, perhaps, in Mauritius. A method
for dealing with the rabbit and goat situation was, we were
informed, in hand.

I have collected animals in a great many parts of the world and
none of the captures has been easy. However, on Round Island the
reptiles went out of their way to be cooperative to an extraordinary degree. Of the lizards, the Telfair's skinks were so tame that
when we squatted down to have a picnic in what little shade
was provided by a pandanus' green hand-like leaves, they joined us

with whole-hearted enthusiasm. Large smooth-scaled reptiles, a sort of greyish caramel colour which was iridescent as a rainbow when the sun caught them, they had pointed, intelligent faces and thick black tongues. They clustered around us, climbing into our laps and partaking of hard-boiled eggs, tomatoes and passion fruit in the most genteel way and sipping beer and Coca-Cola out of our glasses with all the decorum of a group of village ladies at a vicarage tea party. We felt like cads and bounders when, at the end of the repast, we simply picked up our well-behaved guests and bundled them head first into soft cloth bags, rather as the Mad Hatter and the March Hare bundled the Dormouse into the teapot in *Alice*.

Our next objective was the capture of the Gunther's gecko, an eight-inch plump lizard with huge golden eyes, fan-shaped suckers on its toes and a mottled black and ash-grey skin as soft as velvet. They were not as convivial as the skinks and preferred to live in the remaining palm savannah, clinging to the trunks of the trees about halfway up. We had to use a more complex method of capture; we had to to fish for them. We had brought bamboo poles with us and we attached slip knots of fine nylon to these. The geckos were most cooperative, staying quite still until we had the noose over their heads and round their fat necks. Then it was merely a question of chivvying them down the trunk and into a bag. This had to be done very gently and with great care, for should the lizard suddenly panic and pull against the noose there was a danger of the nylon cutting into the tissue-paper-soft skin of the neck. We were successful, however, and we now had twenty skinks and sixteen geckos in the bag – quite literally.

Next came the snakes. The two species found on Round Island are non-poisonous and distantly related to the family of snakes which includes the boa constrictors of tropical America, but are now considered to comprise a family on their own. One of them is an olive-coloured snake with lighter markings, about three and a half feet long. During the day it rests in the skirt of dead fronds which hang down bases of the Latania palms, of which there are quite a few left in the savannah area. These snakes were easy enough to catch since they just lay there and let you pluck them out of the fronds, but the difficulty was that they lay so still they were difficult to spot. With the other species of snake, we had no success at all. It is subterranean and so much more difficult to locate. The last one had been seen in 1975; since then there had been no reports of it, and it was feared extinct. Though we hunted high and low, we could find no trace of it and regretfully decided that it must be so.

When we returned to Mauritius, we found that all hell had broken loose over the rabbit and goat extermination plans for Round Island. The authorities had been advised that poisoning was the only solution since the terrain was too difficult to employ any other method, and so strychnine had been chosen. This is an unpleasant poison but, unfortunately, the only one apparently available at the time which could be left out in the blistering sun for days on end without losing its potency. In hindsight, of course, the choice of strychnine was even more unfortunate, for it was discovered later that it could have poisoned some of the reptiles, as well as the target goats and rabbits. However, strychnine was though the best choice at the time, and it was imperative that Round Island be cleared of the offensive herbivores as soon as possible. Then someone connected with the project gave a press interview in which he blandly outlined the plan and pandemonium erupted.

Various animal protection societies in the UK took up the cause, baying like hysterical hounds, one of their representatives even going so far as to say to Sir Peter Scott (who was attempting to mediate) that he would rather see all the species on Round Island become extinct than one rabbit killed. The local Society for the Prevention of Cruelty to Animals in Mauritius who up till then had been most helpful and had agreed to the poisoning campaign now got cold feet, back-pedalled madly, and said it would be rank cruelty to poison the rabbits and goats and they would certainly do everything to stop it. In vain did we plead that at this rate the rabbits and goats would eradicate their food supply and thus die a slow death by starvation, surely worse than a quick one by poisoning. One English society for the welfare of animals did send out a marksman who spent some time on the island trying to eliminate the goats with the aid of a rifle and, to our considerable surprise, succeeded. I say to our surprise, for the animals were extremely wary and had made their headquarters at the lip of the volcano crater, the most difficult and dangerous terrain on the whole island. But this still left the insidious munching rabbits, and so matters rested.

It was during this extremely worrying time that an amusing incident occurred which lightened our load a little. The man in England who was causing the most trouble over the whole business of rabbit eradication was a certain Dr Glenfiddis Balmoral. This is obviously not his real name, but his name was sufficiently unusual to make it memorable. John Hartley and I, together with my friend Wahab Owadally, Chief Conservator of Forests for Mauritius, were attending a conference at London Zoo and, glancing casually through the list of participants, I saw the dreaded doctor was going to take part. I felt sure that if the three of us could get him somewhere privately we stood a chance of talking some sense into him. We planned our kidnap with great care. Firstly, I asked Michael Brambell, then Curator of Mammals at London Zoo, if we could borrow his house, which stood on the banks of the Regent's Park Canal, not far from the conference hall. Then I got an exalted personage to introduce me to the doctor. He seemed a nice, sensible man and I was amazed that he was taking such an extreme view over the so-called Bunny Blood Bath. I said there was something I and my colleagues would like to discuss with him and would he join us during a break in the conference to have a drink at Michael's house; he agreed with alacrity. Unfortunately he chose the moment when someone was giving a paper on breeding manatees, animals for which I have a passion, but I felt I would have to forgo it as Round Island was so important. It is one of the many sacrifices I have made in the name of conservation.

Anyway, we dragged the good doctor off to Michael's house, raided the drinks cabinet and soon had our victim mellowing under the influence of a huge gin and tonic. I started in on the Round Island problem and its global importance. The good doctor listened, nodding wisely. When I flagged, Wahab took up the reins and explained how the rabbits were doomed to slow death by starvation. He painted such a horrific picture that there were tears in all our eyes. Then John leapt into the fray and explained eruditely why Round Island was of such biological significance that to let it be destroyed by rabbits would be criminal. Throughout this, the good doctor had nodded agreement and said encouraging things like 'Quite right – I agree – yes, very true', so it came as something of a shock finally, when we had run out of steam, to hear him say, 'But I don't really see how I can help you.'

I looked at him blankly.

'But you're Dr Glenfiddis Balmoral, aren't you?'

'Yes,' he said, puzzled.

'Of the Society for the Greater Protection of Fur and Feather?'

'No, no,' he corrected, 'of the Society for the Preservation and Better Understanding of the Coleoptera.'

It was the wrong doctor. But who would have thought there could be two doctors with the same unusual name? The really bitter part of it was that I had missed my manatee paper.

In the meantime, seeds from the rare Round Island palms had been collected and successfully reared at the Botanical Gardens in Mauritius and we were having spectacular success in breeding all the reptile species we had brought back to Jersey. Of the Gunther's gecko we have bred 235, of the friendly Telfair's skink 327 and – probably our greatest achievement – 31 Round Island boas. We have sent specimens of the skinks and geckos on breeding loan to the USA, Germany, France, the UK, Holland and Canada, thus making sure that they were well established in captivity. Some of the boas we sent, appropriately enough, to Canada, to The Reptile Breeding Foundation belonging to Geoff Gaherty, whose enormous gift enabled us to build our own Reptile House.

We now had both palms and reptiles that could be returned to Round Island and all that stood in the way was this bunch of unattractive invaders. We had the 'there' to put native plants and animals back into, but the 'there' was not as yet suitable.

Ever since I visited New Zealand many years ago, I have kept in touch with the New Zealand Wildlife Service, probably the best in the world. In one of my letters, I mentioned the problems we were having with Round Island and asked if they had any suggestions, for I knew they had had terrible problems with rats and feral cats on their offshore islands. Back came the answer from Don Merton who said he thought could solve our problem for us. They had evolved a new form of poison which was specific to mammals and painless, unlike strychnine, and which moreover remained stable in extremes of temperature. In addition, said Don, he felt sure that if we approached the Wildlife Service he and some colleagues would be given leave of absence and would be delighted to place their expertise at our disposal and do the job for us. This seemed almost too good to be true, but in due course Don and his friends arrived in Mauritius carrying goodness knows how many hundredweight of deadly poison in their luggage, as well as tents, tarpaulins and other vital equipment. It was going to be a long job, we knew, for the task of poisoning all the rabbits in 350 acres of terrain which resembled the surface of the moon was not going to be easy. All this heavy equipment would have to be carried to Round Island by helicopter, since it would be impossible to land it all from boats and then carry it up almost perpendicular cliffs. Just as they were all packed up and ready to go on their vital mission, the government helicopter we were relying on broke down. We were in despair at the thought that we might have to put it off for yet another time, but a kindly Fate stepped in to our rescue. A destroyer of the Australian Navy was paying a courtesy call to Mauritius and had on board a spanking new helicopter. Frantic phone calls to the Australian Ambassador met with a kindly response and the Australian Navy was pressed into service to help us. Don and his team, plus their gear and poison, plus a six weeks' water supply (for there was none on the island), were taken by HMAS *Canberra* to Round Island and there the helicopter lifted the whole cargo, both human and inanimate, on to the only bit of Round Island that could be called flat. Here Don and his team made camp and started work.

They did the most wonderful job under the most trying conditions and within six weeks Don reckoned that they had eliminated all the rabbits. However, to make absolutely sure, he and his team paid a return visit a year later. The reason we had to make sure of course was that if there were half a dozen rabbits of both sexes left, the poisoning would have merely acted like a cull, the surviving rabbits would have undergone a population explosion and we would be back to square one. However, there was not a trace of a rabbit and Don sent me some exciting photographs that showed the carunculated, desiccated, eroded surface of Round Island already wearing a green haze of new growth. The island was having its second chance to survive. The work does not stop there, of course, for palms and, we hope, hardwoods will be planted. But we also hope that in fifty years' time the island will closely resemble what it was some two hundred years ago, and offer a safe habitat for its strange and unique denizens.

Just recently, we had a visit from the distinguished author, Richard Adams, who wrote that extraordinary best-selling book about rabbits called *Watership Down*. Jeremy was showing him around the collection and, as Jeremy always does, was getting more and more enthusiastic about our work and particularly about the far-reaching ramifications of it. He was busily telling Mr Adams probably more than he wanted to know about our work on Round Island when he made a mistake.

'Yes,' said Jeremy with enthusiasm, 'and after the goats were eradicated, we managed to get rid of three thousand rab . . . rab . . . rab . . .' Jeremy's voice faltered and came to a stop. How could you tell the man who had written *Watership Down* that you had exterminated 3000 rabbits, without earning a certain amount of displeasure?

They looked at each other in silence for a moment. Jeremy got redder and redder.

'It's quite all right,' said Richard Adams, placidly. 'I don't know why everybody thinks I like rabbits so much just because I wrote a book about them.'

So, in collaboration with the Mauritius government and with the aid of the New Zealand Wildlife Service and the Australian Navy, the Trust had saved Round Island. We had, as I wrote to Roger Payne, saved a 'there' to put 'them' back into. But more than that, I believe that by this achievement we have taken the conception of a 'zoo', as it is popularly thought of, a stage further. Not only have we demonstrated that captive breeding, be it in Jersey or the country of origin of the species, is of vital importance, but we have shown how a 'zoo' can help in the resurrection and protection of the habitat of the animals it deals with.

What we have done for Round Island will, I am convinced, be used as a model for many parts of the world where fragile eco-systems exist and are under the same threat. So we hope we have shown that the zoological garden will have progressed from being the sterile Victorian menagerie (of which there are still far too many examples), to being a vital force in the conservation of the other forms of animal life which share the world with us. We feel this is what all zoos, especially the ones in richer countries, should be doing, and if for financial reasons they cannot venture as far as the Mascarenes they will, if they look, inevitably find on their doorstep a Round Island they can help, such is the urgent need for conservation in the world today.

6

A Festival of Animals

So 1984 became a year of dual birthdays for it was twenty-five years since I had founded the zoo and twenty-one since it had undergone its metamorphosis and become the Jersey Wildlife Preservation Trust. It thus behoved us, in the warm flush of celebration, to take a cautionary look at our progress to date.

What had we accomplished? Well, first, I think we had proved that a zoo can and should be a vital cog in the conservation machine. If captive breeding was mentioned twenty-five years ago within the hearing of a group of earnest conservationists, they flinched and spoke loudly of other things, rather as if you had the bad taste to confess that you thought necrophilia a suitable means of birth control. But three years ago the IUCN issued a policy statement which embraced captive breeding as a vital conservation tool. As a document it makes for interesting reading, for it sets out, almost word for word, the guidelines for captive breeding that we have been using for a quarter of a century, and which I had been preaching since I was sixteen! We are delighted that the conservation establishment has made it respectable at last, but why did it take so long? However, one should not carp. It is pleasant to receive into the fold such exalted disciples, however belatedly.

Thus we have set up breeding groups of endangered species not only in Jersey and various other suitable zoological collections in Europe and America, but also in the countries of the animals' origin. There are breeding facilities designed and built with our help in such places as Brazil, St Lucia, Mauritius, Morocco and

Madagascar. This latter, highly successful *in situ* breeding colony is an interesting departure for us, for it involves the world's rarest tortoise, the Ploughshare tortoise, or Angonoka, a species which we do not have in the collection in Jersey. It may be represented in years to come, but for the moment the important thing is to build up its numbers, which our Conservation Field Officer in Madagascar, Don Reid, is doing with great success. Of course we have not undertaken this project single-handed, but with the help of many organizations, notably the Worldwide Fund for Nature, but Lee is master-minding the whole endeavour and at the present rate of progress it seems set fair to be a wonderful example of how such projects should be planned and run.

The next important thing we have done is to create our mini-university: our International Training Centre for Conservation and Captive Breeding of Endangered Species, to give it its full title, where people from developing countries come on scholarships and return home to run the breeding programmes we have established in accord with their governments.

Just recently, Jeremy came back from an IUCN meeting in Costa Rica and was delighted to have met up with twenty-two of our graduates there, from countries as far apart as Thailand and Brazil, all anxious for news of progress in Jersey – their Alma Mater, so to speak – and eagerly sharing their news and views with one another. Jeremy said it was like a big family reunion and it was most heartwarming to see that the Trust had trained and enthused so many young people from so many different parts of the world.

The Training Centre is now affiliated with the University of Kent in the UK, which offers our trainees the opportunity to study for a Diploma in Endangered Species Management, the first of its kind. We have access to the university's sophisticated computer facilities and can thus speedily and efficiently garner knowledge for our purposes from all over the world. Furthermore, in 1989, the university set up a brand new institute in the field of biological science, and I was greatly honoured when they asked me if they could use my name in this connection. So this institute – the first of its kind in the United Kingdom – is to be called the Durrell Institute of Conservation and Ecology, or DICE for short. I think this is an appropriate acronym since, as I have shown in this book, conservation is a dicey game at best.

Our other activities, both Jersey-based and further afield, have kept pace with our main developments. The results of our scientific research – both on the animal collection in Jersey (and its myriad aspects of veterinary medicine, nutrition and breeding biology) and on the behaviour and ecology of species in the wild – are available in libraries all over the world. Our educational efforts are expanding rapidly, whether they are for Jersey schoolchildren or for the people of Madagascar who see posters of lemurs and Ploughshare tortoises in schools and public buildings everywhere.

Our work has been greatly aided by our members worldwide, but particularly those in North America. As I have related, I went to the United States in 1973 and, with the tremendous help and support of Tom Lovejoy and my other American friends, Wildlife Preservation Trust International was founded. In 1986, Simon Hicks was instrumental in forming Wildlife Preservation Trust Canada. Apart from widening our scope, the creation of these two sisters organizations enables our American and Canadian supporters to get tax relief on their membership dues and their generous donations, both of which support the international conservation efforts of the Trust as a whole.

Finally, we are now starting to see the rewards of our work: the successful release to the wild of captive-bred animals. Such rare creatures as Pink pigeons and Golden lion tamarins have settled down to their new lives in the forests and, most importantly, have begun breeding. It is a wonderful feeling after all the years of carefully building up our breeding colonies to be able to return animals to where they belong and see them flourish.

So all the objectives we set ourselves when the Trust was established have been tackled, some more thoroughly than others, but at least the mechanism is in place and is ready for further development. It is, however, nice to receive compliments and one of the nicest we have received came from Dr Warren Iliff, director of Dallas Zoo and past president of the American Association of Zoological Parks and Aquaria, when he spoke in 1988 at the Southern Methodist University in Texas. He said: 'If you ask people which is the best zoo in the world some say San Diego, some say Bronx. But if you ask zoo people themselves, people professionally involved with zoos, including zoo directors, they say the Jersey Zoo.' A statement like this from the other side of the Atlantic where they have mega-zoos run on mega-bucks is a compliment to cherish.

We have been lucky too, as I have said, that I have written books which have been popular and which have helped us obtain membership, have shown people the importance of captive breeding and have opened many doors and given me access to people I would not otherwise have met and this was made very obvious when we were planning our anniversary celebrations.

Because we are a small organization, the 'boys' (as I insist on calling Jeremy, John, Simon and Tony, to their disgust) frequently come up to our flat at the end of the day to partake of a drink. They do this during the day as well, should we have weighty matters to argue and discuss before taking our recommendations to the Board of Management or to Council. In this way we separate a lot of wheat from the chaff and save innumerable hours of debate at committee meetings. If I am cooking – an art which I enjoy practising – the boys sit around the kitchen table; if not, we repair to the living room where half of them have to sit on the floor, and papers and notes are strewn about in what looks like total confusion but is really very orderly.

On special occasions, such as the birth of a baby gorilla, or the falling in love of a pair of Golden lion tamarins, we drink champagne – not, I hasten to add, paid for out of Trust funds but from my own cellar. This particular morning we felt warranted the popping of corks for we had just heard that Princess Anne would come over to Jersey to join in our birthday celebrations. Now all we had to do was to plan the shape of the affair.

'The princess has consented to open the Training Centre,' said Simon, lying on the carpet clutching his champagne goblet, 'so that's excellent. That can happen first.' Although our mini-university had been operational for a couple of years, it had never been 'officially' christened and naturally we wanted out patron to do this.

'And then what?' I asked, for the bulk of the arrangements would fall to Simon's lot.

'A luncheon,' he said. 'Very select. Trust members only.'

'Speeches?' asked John.

'I hope Gerry will make one and the princess will reply,' said Simon.

'Oh, God, you know I hate making speeches, Si; do I have to?'

'Obligatory,' said Simon. 'Can't have the princess making a speech and the founder sitting there silent.'

'Just a few simple words,' John added encouragingly.

'Good if they're simple you can write them for me,' I said.

'You always change everything I write for you,' said John, indignantly.

'That's because you can't write,' I said. 'Go on, Simon, what about the evening?'

'I've had a brilliant idea about that,' enthused Simon, his blue eyes flashing. We all groaned and Jeremy closed his eyes and a spasm of pain flashed across his face, making him look even more like the Duke of Wellington – after suffering a defeat. We all knew Simon's brilliant ideas.

'What I suggest is,' Simon went on, oblivious of our unanimous horror, 'we hire Gorey Castle and have a pageant there.'

This surpassed all Simon's brilliant ideas. Gorey Castle built in the thirteenth century, dominates the pretty little fishing port which lies below it in a half-moon bay. A magnificent pile of masonry, it seems newly minted, its walls and towers and battlements un-pocked by cannon shot. It looks as though it has just been built by Hollywood and when it is floodlit you expect to see Errol Flynn swagger out on to the battlements at any moment. The castle is the most spectacular on Jersey and had been under the care of Sir Walter Raleigh when he was governor of the island in 1600. The idea of renting such a desirable property was very compelling but, I realized sadly, unrealistic.

'Rent Gorey Castle,' said Tony, scandalized, 'they won't rent you a *castle*!'

'Well, if they know who it's for they'll probably give it to us free,' said Simon, dismissing this quibble. 'Then I thought that this pageant could be mediaeval. We can accommodate about two thousand people, I should think. We'll all dress in period costume and we'll have an ox roasting on a spit and we'll have . . .'

'Two thousand people,' exclaimed Lee. 'Who's going to serve them?'

'Waiters,' said Simon, surprised that Lee had not thought of this simple answer herself.

'Where are you going to get them? There aren't enough on the island as it is,' Jeremy pointed out.

'We'll fly them in,' said Simon, intoxicated with the idea, 'just fly them in.'

'But where will they sleep?' asked Jeremy, exasperatedly.

'Tents,' said Simon, 'we'll erect tents in the castle grounds.'

I had a wonderful mental vision of a host of disgruntled Portuguese and Spanish waiters, dressed in Elizabethan ruffs and plumed hats, crawling in and out of tents in the pouring rain.

'What about toilet facilities?' asked Tony who, as our general administrator and veterinarian, had spent a very unpleasant few hours the previous day sorting out a problem with our ladies' lavatory and so was inclined to a morbid view.

'Dig latrines,' said Simon promptly.

'Who will dig them?' asked Jeremy.

'Volunteers,' said Simon.

'And if you can't find volunteers?' asked Jeremy.

'Ask the waiters to dig them,' suggested John.

'And where are you going to get an ox?' asked Tony, the veterinary side of his nature coming to the fore.

'Buy one,' said Simon.

'Public health would never allow latrines all over the castle,' said Jeremy.

'Apart from the hygiene – the smell,' Tony added with feeling.

'Better a banquet where herbs are rather than a burnt ox,' I said, misquoting the Bible.

'We'll have a banquet too,' said Simon, clinging tenaciously to his idea, 'venison and things like that.'

'We could melt some lead and pour it from the battlements over those people we disapproved of,' John suggested, helpfully.

'Lead's extremely expensive,' said Jeremy, taking the suggestion seriously.

I felt the meeting was getting out of hand, so I opened another bottle of champagne.

'Look,' I said, 'fascinating though this castle idea is, it is bristling with pitfalls and I have no wish to have to explain to the Palace why I entertained the princess in a castle lashed by wind and rain, with half-cooked oxen all over the place, boiling lead dripping from the battlements and a lot of foreign waiters complaining that their cod-pieces didn't do them justice, or were uncomfy.'

'You mean you don't like my idea?' asked Simon, crestfallen.

'An excellent idea, but for some other occasion,' I said. 'But *I* have an idea. How about having a sort of celebration of animals and inviting over all the celebrities I know who are connected with conservation so that each can show the importance of animal life to them?'

Simon brightened. 'You mean, a sort of stage show?' he asked tentatively. I could see the fanatical gleam returning to his eye.

'Well, yes,' I said, vaguely. 'I mean actors and actresses to read poetry about animals, a ballet perhaps, get Yehudi Menuhin to play something like the 'Carnival of the Animals' – that sort of thing.'

'Yes, yes, jolly good,' said Simon, gazing into space as if he could see it all. 'We'll do it down in St Helier at Fort Regent. They have a huge stage and all the equipment, lights, a huge projector for film, quadraphonic sound. Oh yes, it'll be absolutely splendid. Jolly good.'

Thus was the 'Festival of Animals' born. The cast list was impressive and interesting as I had met most of these celebrities in so many different ways.

For reading the bits of poetry, I wanted two contrasting voices, male and female. Of all the excellent actors I knew there was one whose voice stood out: Sir Michael Hordern. When he spoke, it was like listening to a wonderful vintage port which had been given the power of speech: rich, resonant and refulgent. I did not know him, but I knew that he had read my books and liked them, so I was delighted when he accepted. About the choice of the female voice I was never in doubt. Ever since I had first seen her in that enchanting film, *Genevieve*, about the London-to-Brighton vintage car race, I had fallen deeply and irrevocably in love with Dinah Sheridan. Later I had seen her in the film *Where No Vultures Fly* and my heart was even more firmly entangled. However, I knew that she was married and, upright and honourable as I am, this prevented me from searching her out and laying my heart at her feet. Another reason was, of course, that I was married myself. So I had rconciled myself, not without a struggle, to life without Dinah Sheridan.

Then just before our great anniversary, two things happened. I was going to London and I saw they were reviving *Present Laughter*, a very funny play by my old friend Noël Coward, who had for some years been one of our overseas trustees. In the cast, to my delight, was Dinah Sheridan; so I determined to go to see my paragon in the flesh, as it were. As we were flying to London, I was idly leafing through the in-flight magazine and came upon an interview with Miss Sheridan. Among the normal spate of vacuous questions one is inevitably asked at such interviews, she had been asked who she would like as a companion if she were wrecked on a desert island. She had answered 'Gerald Durrell'. I could not believe my eyes.

'Good God, she wants to be wrecked on a desert island with me,' I said to Lee.

'Who does?' asked my wife suspiciously.

'Dinah Sheridan.'

'What would she want that for?' asked Lee, in the dampening way wives have.

'Because I'm a fine, upstanding fellow of high moral principles,' I said.

'If she said that, you can tell she's never met you,' said Lee crushingly.

But I remained uncrushed. I was aglow. She might well have chosen that cad Attenborough, or that bounder Peter Scott, but no, she had chosen me. So a dozen carefully chosen yellow roses, unblemished by greenfly, black fly, earwigs, death-watch beetles or similar pests were purchased and sent around to the stage door with a card saying, 'The one clapping the loudest will be me. Can I see you after the performance?' Back came the answer 'Yes'.

The wit of Coward combined with a glittering performance by Dinah made an unforgettable evening. Later, drinking Scotch in her dressing room, I confessed my adoration of many years' standing, and we decided to meet frequently in spite of Lee. When she related this to her husband Jack, he sent me a stiff note accusing me of alienating his wife's affections with over-fulsome adulation and yellow roses, and challenged me to a duel at dawn in Hyde Park. I accepted, but pointed out that as he had challenged me mine was the choice of weapons. I suggested champagne corks at fifty paces. It was on this happy note that our friendship began, and so when we were looking for the actress for our Festival of Animals, Dinah was the obvious choice.

I had first met Yehudi Menuhin in France when he had come to stay with my elder brother Larry. Our little house lies some twenty-five miles away from the village where my brother lives (twenty-five miles is the requisite distance to keep between yourself and an elder brother) and so Lee and I drove over to have lunch with Larry and the Menuhins, an occasion which was joyous because Yehudi and his wife were charming. Lunch was lengthy, with ample food and wine, and about four o'clock we all began to think wistfully of bed and the word 'siesta' began to be mumbled. Fortunately, Larry's house is huge, with an amplitude of bedrooms, so Lee and I chose one and were soon asleep. When we awoke we heard the sound of a violin.

'Who's playing the gramophone?' asked Lee.

'That's Yehudi practising,' I said.

We crept out on to the landing and from a bedroom not far away came the sweet song of the violin, played by a master. I have in my life been woken from a siesta by many sounds – the song of birds, the crash of thunder, the purr of a stream and the silken sound of a waterfall – but never had I been woken as beautifully as that.

Of course we asked the Menuhins and Larry to lunch with us the following day and, having discovered that Yehudi liked lentils, rice, peas and such things, I created a special Menuhin curry of colossal proportions. We were going to eat at our long table on our patio. There had to be innumerable side dishes and, to save time when laying the table, Lee had carefully arranged all the spoons, knives, forks, ladles and so on in special order on a large tray. Our guests arrived and, after a few drinks, Lee went off to the kitchen to put the finishing touches to the lunch. After a short while, Yehudi followed her and surveyed her, busy in the kitchen.

'Do let me help,' he said and, without waiting for Lee's reply, his eye fell on the tray of cutlery which he seized and carried out to the patio before she could stop him. Beaming, he approached the table and emptied the carefully arranged trayful of cutlery on to it in a great, gleaming, clattering, tangled pile. I saw Lee's expression of horror, so I shepherded Yehudi back to the chairs, gave him another drink and went to help my distraught wife disentangle the feeding instruments.

'I spent so much *time* on this,' she whispered.

'Never mind. Look on the bright side. It's not every hostess who can say she has had her table arranged by Yehudi Menuhin,' I pointed out.

So I wrote to Yehudi and he, generous and kindly man that he is, said he would be glad to give his services to our cause and play something with the Jersey Youth Orchestra.

So now we had a famous actor and actress to read poetry, an orchestra and a violinist of renown. But there were still many other ways in which the animal world touches and enriches our lives that I wanted to illustrate. In dance, for example; in song, in television and in painting. A friend of ours, Jeremy James Taylor, who had agreed to produce the show, had contacts with the Royal Ballet and, to my delight, they agreed to send over a band of up-and-coming students to perform for us.

A few years previously, I had done a spot on a children's television show with vivacious Isla St Clair. During rehearsals she had talked a lot about our activities and had seemed deeply interested in what we were doing, so interested indeed that she made a fatal mistake. She said that if, at any time in the future, I needed her help I had only to let her know. Having heard her sweet and lovely voice and been enchanted with it, I felt she would be the ideal person to represent animals in song. I phoned her, reminded her of her promise and asked her to come to Jersey. She said she would be delighted and, in fact, knew an attractive little song about a zoo.

I then had a pause for thought. What, I said to myself, about plants? After all, without plants animals could not survive. Of course, there was only one roaring, uninhibited champion of the plant world and that was David Bellamy. Then a wicked thought crossed my mind. Flanders and Swann, in their brilliant two-man shows, *At the Drop of a Hat*, had done a song called 'Misalliance'. It was about a honeysuckle and a bindweed who fall in love but, because one turns clockwise and the other anti-clockwise they can never marry and so 'they pulled up their roots and just withered away'. By dint of bribery and corruption I got Isla and David to sing this as a duet – a most unlikely combination, as David has –

and I am sure he will not be insulted by this – a voice like the mating call of a romantic bull walrus.

The choice for animals on television was easy, for who could be better than David Attenborough?

I had known David since he was a lowly BBC producer. We were introduced in a pub, where we spent a convivial morning discussing animals and travel. Some years after this, David phoned and asked if I could do a radio programme with him in the zoo. I said I would be delighted, and so a date was fixed.

In those days we still possessed Chumley and Lulu, our pair of chimpanzees of uncertain virtue. When you went to visit him, Chumley, after his hysterical morning greeting, which consisted of bared teeth, loud maniacal screams and swinging to and fro around the cage, would sit down and dissect an orange with the deep concentration and delicacy of a famous Harley Street specialist performing a lobotomy on a Prime Minister. Lulu, well aware of her husband's impeccable manners when it came to the weaker sex, took no chances; while her husband was busy with his display she stuffed her mouth full of grapes, gathered together as much fruit as she could and sat on it in the hope that it would escape the attention of her spouse. Chumley, having completed his surgery on the orange, ate the contents and threw the skin at Lulu, generally hitting her on the back of the head. Chumley was an underarm bowler but his skill and accuracy were remarkable. Having thus informed Lulu of his devotion, he leapt on her when she least expected it, cuffed her over the back of the head and dragged her, screaming, off the pile of fruit she was hiding. He then sat down, stuffed a banana into his mouth, masticated it into a suitably pulpy condition, and spat it into his hand to investigate it with a fat forefinger, like somebody sorting out change for a vending machine.

The one thing you could always rely on Chumley to do was to cause you the maximum amount of embarrassment. Take a party of exalted visitors around and Chumley was ready for you. He seemed to know that the people were important and that he was supposed to be on his best behaviour. A malevolent gleam would appear in his eye as he summed up the situation and thought out the best strategy for causing havoc. He generally began by beating up Lulu, pulling her hair or knocking her over and jumping up and down on her. He did this for two reasons. First, Lulu had the loudest and most penetrating scream of any chimp I have ever heard: a cross between a demented train whistle and a knife blade on a plate. Second, he had discovered that there was nothing quite like a little domestic upheaval to focus the interest of his audience. Having assured himself of their undivided attention, he then continued his act by raping his wife or else he sat on a branch doing unmentionable things to himself with great gusto, so that the ladies in the group would go quite pink and fan themselves with their guide books. Then, when everyone had been more or less lulled into a sense of false security by this temporary immobility, he would regurgitate a great handful of masticated fruit and scatter this largesse over the crowd, who would run screaming from the cage with gobbets of glutinous fruit adhering to their clothes. To force a crowd to break and scatter like this made Chumley feel good: he knew he had achieved the height of his ambition and life could offer nothing more exquisitely pleasurable.

Despite the many years I have spent taking important people round the zoo, I always approached Chumley's cage with a feeling of extreme trepidation, which always proved to be well-founded. So I remember vividly the day that David Attenborough came over to do the radio programme with me.

It was a very simple programme, with David and me ambling from cage to cage telling anecdotes about the animals we had met in various parts of the world. It was not the sort of programme you could get away with now, for your audience demands technicolour and huge close-ups of the animals you are talking about. But in those happy, far-off days of steam radio audiences were less demanding. The first thing was to take David around the collection, so that we could decide which species to have in the programme and who was going to say what about which. It was his first visit and although we were still in our primitive stages he was infectiously enthusiastic about our animals and our aims. We were enjoying ourselves so much that we approached Chumley's abode without a tremor of doubt sullying my mind. As soon as David saw the apes, he uttered a delighted cry and hurried to the front of the cage. It just so happened that in that particular week we had been inundated by a flood of exotic fruits. The chimps had engulfed their share of this bounty with vociferous enjoyment, but it had brought havoc and disaster to their bowels. In consequence, to say that the cage was well stocked with throwable material of the more adhesive kind is an understatement. Chumley, seeing his victim approach in all innocence, was delighted. He scooped up two great handfuls of ammunition and, as David reached the barrier rail, released them with unerring accuracy. They hit David amidships, as it were, and his immaculate white shirt became a midden. David stood aghast while Chumley, encouraged by his success, loosed off another two handfuls which hit their target with equal accuracy. David was beginning to resemble a walking manure heap when I rescued him. Apologizing profusely, I whisked him into the manor, where he could wash, and lent him a clean shirt. After a very large drink he seemed slightly mollified, but for the rest of our tour I noted that he approached all cages with circumspection and, where possible, allowed me to precede him.

Hoping the years had dimmed his memory, I phoned David, reminded him that he had never returned my shirt, and said I would like him to come and show that enchanting sequence from *Life on Earth* where he sat surrounded by Mountain gorillas in the forest, one of the most moving sequences in the series. To clinch it, I said that our mutual friend of long standing, Chris Parsons (who had been boss man of the BBC Natural History Unit when *Life on Earth* was made) had agreed to come over to handle the technical problems of projection and so on. David agreed at once and was as good as his word, even though he did not bring me a replacement for my shirt.

As with animals in television, animals in art presented no problems, for who better to illustrate this than David Shepherd? As a

superb painter, David had long since lost his heart to the wildlife
of Africa, particularly the elephant. His vivid and magical paint-
ings of pachyderms and other beasts had earned him a worldwide
following and he used the money he earned from his work to
create his own foundation for the protection of African wildlife.
Before I met him for the first time, I was told that I was bound to
get on with him as we were each as mad as the other. When we met,
I grant you there were certain similarities but I still maintain that
David has the edge over me, for I would not be so idiotic as to go
palette in hand, trailing a BBC film crew, in an effort to paint an
original portrait of an elephant in the wild. I forget how many
times they were charged in this ridiculous and dangerous process
but I know Chris Parsons, who produced it, came back from Africa
with grey streaks in his hair and a haunted look in his eyes. However, David agreed to come along and show the piece of film of
him being pursued by an elephant and to talk about the importance
of wildlife generally and its importance in art.

Johnny Morris had for years represented the rather gormless
Victorian-type zoo keeper in a series on TV called *Animal Magic*. I
had known him for years – a gentle, kindly man whose powers of
mimicry were extensive and incredibly funny. Once, when he was
making a film in the Scilly Islands, I had lent him my dog (also
called Johnny) as a prop-cum-mini-star and so I felt that one good
turn deserved another. Johnny said he would be delighted to come
and tell a funny story of him (in his keeper role) having trouble
with an elephant.

So now everything was more or less in place, but still Simon
was running around like one demented. Hotels had to be booked,
flowers put in rooms and a host of other details attended to. His
job was not made any easier by the harsh truths of meteorology.
We knew, from bitter experience, that if there was a gale anywhere
in the Atlantic (say down near the Falklands) it would invariably
turn up at Jersey to discomfort us. If there was so much as a wisp of
fog anywhere between the South Polar ice-cap and the English
Channel, it would make its way with unerring accuracy to settle on

Jersey like an opaque teacosy, so that planes could not get in or out. This generally happened when we had been entertaining a bore we were desperate to get rid of, or looking forward to the arrival of a loved friend. On this occasion, it was savage crosswinds, and a few hours before the show David Attenborough and another piece of vital equipment were being blown to and fro over the island, the pilot saying he could not land and Simon telling him he just had to. Finally, they did, before Simon's red hair turned white.

Now he had all the VIPs on the ground and tucked away in various hotels, Simon had to issue them with special passes to allow them access to the giant hall in Fort Regent, for the security surrounding the princess was, naturally, comprehensive. So eager was Simon to make sure he got all these multifarious details right that he omitted to give himself a pass, and while all the stars were allowed into the hall Simon was stopped by the security people. For some time he pleaded that he was the organizer of the show, but they were stone-faced and adamant. He could not enter without a pass. Finally, when he was on the edge of a nervous breakdown, somebody identified him and, with the greatest reluctance, he was allowed in.

At last, the Festival started. I introduced it, explaining that what we were trying to do was to show the importance of the other animals on the planet and how they influenced our lives in so many ways. I was flanked as I spoke by Simon's adorable identical twin daughters (then four years old), both dressed as Dodos in elaborate costumes which I feared might suffocate them under the heat of the lights. Before leading them off, I explained that, as our symbol was the Dodo, it was appropriate that two Dodos were on the stage with me, but unfortunately they were not a breeding pair.

The evening went with a swing and it was obvious that our stars were enjoying themselves as much as the audience was enjoying them. As I sat watching all my friends give so joyously and generously of their many talents, I reviewed the day. It had been long and complicated, and the weather had not been kind. Whenever we are lucky enough to be visited by our patron, the island is immediately lashed by howling gales and rains such as are normally never found outside a serious monsoon. This, of course, meant the last-minute rejigging of events which were supposed to be out of doors and had to take place indoors. But apart from the inclemencies of the weather there had been other and greater horrors which had not been vouchsafed to me.

There was – to choose just one – the case of Motaba's head. You would think that the complexities of a royal visit were enough to cope with without having a gorilla's head inserted into it, but, as I explained at the beginning of this book, when you are living among fifteen hundred animals, practically anything can happen at any time. One learns to live with it; one tends to accept it as a normal pattern of life. But when you get a princess inextricably entwined with a gorilla's head, you feel that life is dealing you a very unfair body blow.

This is what happened, and I am glad that I did not know it while I was taking the princess around: the rain had paused to draw breath while we opened our Training Centre and the princess met some of our multi-coloured, multi-lingual and multi-orientated students from all corners of the earth. After this, it was planned that we went to the manor house so that the princess could sign the visitors' book and then have a tour of the zoo ending up at the gorilla complex at precisely eleven o'clock. Royal visits have to be stop-watched to an exact degree and if we did not reach the gorilla complex at eleven exactly it would throw into disarray all the ensuing events.

Richard Johnstone-Scott, without doubt the best and most experienced ape keeper in the world, had looked at the weather and decided that he was not having the princess view his deeply cherished charges dripping with rain in their spacious outdoor area. They would have to be shown in the inside bedroom area. So, with a flash of genius, Richard created what could be called 'instant jungle'. With branches from our oak trees, our sweet chestnuts and limes, he piled the bedrooms high. The effect was spectacular and when the gorillas were let in they rumbled and growled like volcanoes, as gorillas do when they approve of something.

At that moment, Princess Anne was being ushered into the manor house and the visitors' book was being brought out. We were due at the gorilla complex in four minutes' time.

At this precise moment Motaba decided to get his head wedged between the bars which lined the top of the gorilla bedroom and served two purposes: they kept inquisitive hands from pulling down the ceiling and acted like bars in a gymnasium on which the young apes could swing and dangle and get the maximum amount of exercise. Motaba had found the one place where these bars were fractionally wider apart and, of course, stuck his head through.

Richard was distraught, but no more so than Motaba's parents, Nandi and Jambo. It was an unpleasant affair, even leaving aside the royal visit. Motaba, as all children would in similar circumstances, started to wail and scream, which agitated his parents still further.

We had then a dear and beloved Welsh lady, a Mrs Hayward, who was one of our volunteers. Seeing the pandemonium, she decided that the best thing she could do was to inform Jeremy, as Zoological Director, of Motaba's plight. Jeremy, of course, was with the royal party. Mrs Hayward galloped out of the gorilla complex, down past the Pink pigeons and Palawan peacock pheasants, giving them a nasty shock, raced across the causeway, where the flamingoes viewed her with alarm, sped down the main drive past our giant granite wall, hurled herself through the fifteenth-century granite archway into the forecourt and had just reached the front door of the manor house, panting with emotion, when she was engulfed by a brawny arm and a revolver was pushed into her ribs.

'And where do you think you're going?' asked the security man, benignly.

'To tell Mr Mallinson about the gorilla's head,' she squeaked.

'A likely story,' said the security man.

'But it's true,' she panted, 'the gorilla's got his poor little head jammed in the bars and only Mr Mallinson can save him.'

'Well, they're all very busy in there signing things,' said the secuity man. 'Don't you fret. Stand here until they come out and *then* you can tell them about your gorilla's head.' And, having satisfied himself that she was an unarmed, harmless lunatic, he reholstered his revolver.

Meanwhile, back at the gorilla complex things had gone from bad to worse. Urged on by their offspring's screams, Nandi and Jambo were attempting to free him by pulling him downwards. Richard was terrified that in their well-intentioned efforts they could break Motaba's neck, so he hurried up on to the roof of the complex and tore off a skylight under which Motaba was dangling. Now, all our gorillas adore Richard as much as he adores them, but under stressful conditions apes – like humans – are apt to behave strangely. As Richard tore off the skylight, Jambo's huge, muscular arm and enormous hand flashed up through the bars and Richard fell backwards on to the roof, thus avoiding a blow that would have felled Muhammad Ali, let alone anyone of lesser stature. When Richard got to his feet, Jambo had slid Motaba further along the bars on the roof out of Richard's reach. Richard had to rip off another skylight but was again faced by the protective father. All he could do was to talk calmly and quietly to Jambo, who by then had discovered (sagacious animal that he is) that it was wrong to pull at Motaba and so with one massive hand was supporting his child's bottom.

Through the driving rain, Richard could see the mushroom bed of umbrellas bobbing across the causeway, which indicated that the royal party was almost there. Suddenly, to his astonishment, Motaba was no longer there. Supported by his father's giant hand, he had found the one spot between the bars which had allowed him to push his head through, and now he pulled it out. With great relief, Richard got down from the roof and went to survey the bedrooms. He was horrified.

As is always the case when great apes are under stress, they void

171

an extraordinary amount of excrement and urine, so that Richard's carefully arranged 'instant jungle' was like a manure heap. There was nothing he could do, for the royal party had arrived. When he told me this story the following day, I wondered how I would have coped with the conversation with our patron.

'Oh, yes,' I would have said, 'we always keep our gorillas knee deep in excrement, they seem to prefer it. And that little chap in there dangling from the bars like a hanged man on a gibbet? Well, gorillas frequently do this. It's a sort of . . . sort of *habit* they have. Yes, very curious indeed.'

Fortunately, by Richard's and his friend Jambo's combined intelligence, this dismal conversation was avoided.

Later, at the special luncheon for Trust members I had, of course, to make a speech. It was one of those speeches which, having uttered it, you wished you had not. When the last banal words had been thrust upon my audience, I sank back thankfully in my seat. It was then that our patron took me by surprise. She rose and made a charming and flattering response, and at the end she turned to me and said:

'But in particular I congratulate the man who has made the Jersey Zoo and the Jersey Wildlife Preservation Trust both to be admired and respected worldwide, and as a token of that gratitude I would very much like to present him with a little gift from his zoo staff. I have no need to explain its significance to him, but I am sure we would all like to be included with its thanks and good wishes for the future.'

She then handed me a small velvet bag. On opening it I found a small silver replica of a Bryant & May matchbox. My first thought was: 'Why on earth are they giving me a matchbox when they know I gave up smoking years ago?' Then I opened the matchbox and understood, for inside was a gilt mother scorpion and her babies. It recalled, of course, a scene in my book, *My Family and Other Animals*, when my brother Larry inadvertently opened a matchbox during lunch in which I had earlier incarcerated a scorpion and her young – the pandemonium at the lunch table may be imagined and it made me the most unpopular member of the family. I explained the contents of my matchbox to the audience, most of whom had read the book and were vastly amused at the aptness of the present.

As I sat later and watched the Festival of Animals, I felt in my dinner-jacket pocket the rectangular little shape of my matchbox. I thought what an extraordinarily lucky man I have been to be surrounded by friends who have helped me turn the matchbox zoo of my childhood into a real organization which can and will help the animals which make the world such a fascinating place, a place that we should all cherish.

Envoi

The matchbox of little gilt scorpions was a touching and illustrative gift, for the matchbox of my youth, my boyhood zoo, had become the Trust, one of the world's foremost exponents of captive breeding to aid conservation, with its two sister organizations – one in the USA and one in Canada, and a chain of breeding projects and graduates of its training programme across the world, built up by a devoted and dedicated staff. We had achieved a lot but it was really a drop in the ocean, a leaf in the forest of what still has to be done. I have frequently said that my ambition is to close Jersey down and disband the Trust – because it was no longer necessary. Alas, I fear that day is very far away. But until then I hope we grow and prosper and help preserve the only world we have.

If you have read this book with interest and some amusement, I hope I have managed to show you the complexities and difficulties of making progress in what I believe to be the most important work we humans have to do: the preservation of our planet. If you agree, and would like to join us, we would welcome you with open arms. The more members we have the greater our voice becomes, and in consequence the greater our power for the cause of conservation. We have already built ourselves up from a minuscule project to a world force in conservation, but we have done this only by the support of our worldwide membership. If this book has given you pleasure, has perhaps given you pause for thought, may I ask you to become one of our supporters? We believe that what we are doing is important. We hope you will too, so write to me at:

Jersey Wildlife Preservation Trust
Jersey Zoological Park
Trinity,
Jersey,
Channel Islands,

or

Wildlife Preservation Trust International,
34th Street and Girard Avenue,
Philadelphia,
Pennsylvania 19104
USA,

or

Wildlife Preservation Trust Canada,
219 Front Street East,
Toronto,
Ontario M5A 1E8,
Canada.

(pause, then to captions)

Index

INDEX